Plasticity in the Life Sciences

Plasticity in the Life Sciences

ANTONINE NICOGLOU

The University of Chicago Press
Chicago and London

The University of Chicago Press, Chicago 60637

The University of Chicago Press, Ltd., London

Published 2024

Printed in the United States of America

33 32 31 30 29 28 27 26 25 24   1 2 3 4 5

ISBN-13: 978-0-226-83714-7 (cloth)

ISBN-13: 978-0-226-83716-1 (paper)

ISBN-13: 978-0-226-83715-4 (e-book)

DOI: https://doi.org/10.7208/chicago/9780226837154.001.0001

Library of Congress Cataloging-in-Publication Data

Names: Nicoglou, Antonine, author.
Title: Plasticity in the life sciences / Antonine Nicoglou.
Description: Chicago : The University of Chicago Press, 2024. | Includes
    bibliographical references and index.
Identifiers: LCCN 2024024444 | ISBN 9780226837147 (cloth) | ISBN 9780226837161
    (paperback) | ISBN 9780226837154 (ebook)
Subjects: LCSH: Phenotypic plasticity. | Evolutionary genetics. | Variation (Biology) |
    Genotype-environment interaction.
Classification: LCC QH438.5 .N53 2024 | DDC 572.8/38—dc23/eng/20240620
LC record available at https://lccn.loc.gov/2024024444

♾ This paper meets the requirements of ANSI/NISO Z39.48-1992
(Permanence of Paper).

For Iro

Contents

Introduction   1

Contents

INTRODUCTION

Little by little, step by step, the small child constructs its environment. Similarly, the scientist gradually constructs his reality. Science imitates nature no more than does art. It recreates it. In breaking down what he understands of reality to recompose it in another way, the painter, the poet, or the scientist each builds his vision of the universe. Each fashions his own model of reality in choosing to throw light on aspects of his experience he judges the most telling, and in brushing aside those that are uninteresting to him. We live in a world created by our brains, with continual comings and goings between the real and the imaginary. Perhaps the artist takes a little more of this and the scientist a little more of that. But it is simply a matter of proportion. Not of nature.

FRANÇOIS JACOB (1998, 145)

On a small farm in the Netherlands, a worried farmer watches over a goat as it prepares to give birth to a baby goat. He has been waiting for several hours and hopes that the process will be over soon, fearing for the health of both his goat and the baby. He would have liked the local vet to be there to make sure everything went well, but he was delayed in another village a few miles away, where he was helping a sick cow. The eagerly awaited moment arrives, the baby goat's front legs appear. It should not be long now. A few seconds later, a shapeless mass, still wrapped in its membranes, lies beside the mother. She doesn't move, too exhausted from the effort. The farmer takes action and tries to clean the animal with a bit of straw. Gradually, the little body unfolds. The farmer desperately tries to understand what his eyes cannot see: the baby goat has no back legs, only two front legs. But even before the farmer realizes this monstrosity, the mother begins to lick her little one to bring it to life. Soon she is instinctively pushing him to stand up. The little one struggles clumsily at first, falling, falling again until finally, with the help of its mother, it manages to stand on its front legs for a few seconds, which will become, with time, a few minutes on his front legs, as if in a precarious balance. Thus, it begins its life as a two-legged goat, a life that will be long and fascinating to humans.

In the northwestern Indian Ocean, a few miles from the Arabian Peninsula, a shoal of clown fish has settled near a beautiful sea anemone that protects them from potential predators. But for a few days now, the group has been disoriented: the only female of the group has disappeared. She may have been caught by fishermen or eaten by a shark. Without this female the survival of the group is threatened. But this instability does not last long: the dominant male is already transforming. In a few days it will take the place of the missing female, and there will be nothing to indicate that he was once a male.

In the southwest of France, a butterfly is feeding on the flowers of a white nettle at the edge of a forest. We can recognize it by the upper part of its wings, which is orange and sharply outlined in black and white. It is the middle of spring, and it is hoped that it will find a mate quickly, as it does not have many weeks left to reproduce and ensure a brood. A few months later, in the same area, near a dried-up pond, a magnificent butterfly can be seen perched on a flower. It has beautiful black wings with a white band in the middle and a small dark red line underneath. It is easily recognized by the butterfly hunter because the network of white lines decorating the purple underside of its wings has earned it the name "map butterfly." However, the one with a very different appearance observed in the spring will turn out to be the same species and have the same name even though it does not have a map on its back.

On a tributary of the Adour river in the Dordogne region of France, a carpet of buttercups hides the bottom of the water. The vision is similar to that of a field of buttercups that have accidentally taken up residence in the river current. The flowers point to the surface of the water like so many dots projected onto a dark screen. Here and there, rounded green leaves complete the picture. Like the leaves of a water lily, they seem to float on the surface of the water. A meadow on the water. However, as you approach and look closer, you realize that not only are the flowers firmly attached to the bottom of the river by their stems but that there are many more leaves below the surface. Through the transparent current, you can see them intertwined, like long, smooth, green hair. It is the same plant, but the leaves are different on the surface and under water.

In a laboratory at the University of Maastricht in the Netherlands, a biologist sets out to make steaks from beef muscle stem cells. He places a gel in culture dishes with attachment points that act as tendons for the muscle to develop. He then adds muscle stem cells previously taken from beef. In less than forty-eight hours, what was then just a shapeless puddle in the middle of the culture gel becomes a real muscle, stretched between its two anchor points. Electrical stimulation even allows the various small muscle fibers to regroup to form a homogeneous muscle. In a few weeks' time he has enough material to be able to test the taste of the resulting steak (with or without barbecue sauce).

All the phenomena just described represent moments of pause or astonishment in the description of the living world that surrounds us, moments for which the biologist is led to refer to the concept of plasticity. Whether it is the reorganization of the internal structures at the origin of the bipedal locomotion of the little goat, the transformation of the male clown fish into a female clown fish, the different types of patterns that the map butterfly adopts depending on whether it develops in spring or in summer, the different shapes of the leaves of the water buttercup depending on their position above or below the surface of the water, or the stem cells that divide, multiply, and differentiate into muscle cells, each time the biologist invokes the remarkable "plasticity of the living world." Is the use of this term just a way of marveling at the diversity of living forms? Or is it a way of characterizing a set of phenomena that, if not always as spectacular as in the cases described, are nevertheless general and constant in the life sciences? In other words, is plasticity a cloak used to hide the biologist's ignorance, or is it a scientific concept used to describe an original and specific characteristic of how living organisms interact with the environment in which they live?

Contrary to what a large number of biologists believe and for almost a century have often tried to demonstrate, living beings are not determined by their genes—at least, not in a simple and direct way, as was thought a few decades ago. The milieu, the environment, the nature of the living beings are perhaps subordinate to the importance of genes for their ontogenetic development. Nevertheless, they are essential elements for any change, for any modification of the organism on an individual scale. More fundamentally, many theorists now wonder whether this individual capacity for change might not also play a role in explaining the origin of species diversity, that is, whether it might be a determining factor in evolution.

Starting from the observation of this new transformation at work in the biological sciences, in this book I have tried to understand the reasons that might have led contemporary biologists not only to take an interest in the question of plasticity as a phenomenon that would allow them to move beyond the genetic determinism that had become pervasive in the literature of their predecessors but also, more specifically, to understand the ways in which they used the term *plasticity* in their writings in order to try to understand the phenomenon at a theoretical level. By looking at how embryologists, for example, thought about plasticity and how they used the term, it has become clear that the use of plastic- terms was very often determined by older philosophical ideas. By analyzing in detail these ancient philosophical conceptions—from Aristotle to the beginning of the eighteenth century—I was able to identify the origins of two major trends in the use of the concept of plasticity, traces of which can be found in contemporary biology. The first trend is based on the use of the term in its active sense, as "the ability to generate a form," and the second trend is based on its passive sense, as "the ability to adopt a variety of possible forms."

This return to what I have called "theories of plasticity" (chap. 1) then allowed me to proceed to an epistemological analysis of the concept of plasticity

in embryology (chap. 2), on the one hand, from the seventeenth century until
the end of the twentieth century and in genetics (chap. 3), on the other hand,
from the beginning of the twentieth century until the early 1940s. Comparing
these two histories—the concept of plasticity in embryology and the concept
of plasticity in genetics—has allowed me to show that although the epistemology of the plastic- terms in embryology is still closely linked to the theories of plasticity going back to Aristotle and more specifically to the theories
of generation of the seventeenth century, the epistemological analysis of the
phenomenon of plasticity, as it is perceived and conceptualized in genetics,
opens a new page in the history of the analysis of living beings as the concept
of plasticity gradually becomes an operative concept in this field.

This analysis drove me to consider how the two possible meanings of the
concept of plasticity—the active and the passive meaning—which correspond
to the two main trends in the use of the term that I have just identified—in
embryology and in genetics—came to be articulated within the theory of natural selection at a time when natural selection was gradually being extended
to all areas of biology (chap. 4). I wanted to understand not only how Darwin himself might have contributed, even incidentally, to the modification
of the conception that the biologists of the time had of the phenomenon of
plasticity by associating the term *plasticity* with the notion of *variation* (the
latter notion rapidly becoming central and decisive for the understanding of
evolution by natural selection). Nevertheless, the analysis of Darwin's use of
the concept of plasticity as a "potentiality for variation" has led me to consider that Darwin's original conception of plasticity was hardly imposed in all
areas of biology. One of the reasons for this impasse is the emergence of the
operational concept of "phenotypic plasticity" in genetics at the beginning of
the twentieth century (chap. 5). It is therefore understandable that despite the
repeated presence of the term *plasticity* in Darwin's main works, the concept
is never considered central or of great strategic importance by his commentators themselves.

With the emergence of the operative concept of "phenotypic plasticity,"
the "active" meaning of the concept—which also corresponds to a more
generalized use of the term stemming from the embryological tradition—
disappears completely among twentieth-century evolutionary biologists.
However, the introduction of the concept of plasticity into the field of evolution will raise some controversies within the community of biologists,
particularly regarding the question of the ontological status that should
be attributed to plasticity. Several theorists have wondered whether plasticity should be considered as a trait of organisms among others, a trait
determined by genes (or even by a specific gene) or, on the contrary, as a

by-product of evolution by natural selection. This debate is still very active today and will be discussed even at the risk of perpetuating some of the confusion that surrounds the use of the term plasticity in contemporary biology (chap. 5).

In the second part of the book, I propose an original way of distinguishing between different phenomena described by the term *plasticity*, based on the type of explanation that biologists seek to provide when they refer to the concept of plasticity. Two types of approach are possible. A first approach concerns cases where the concept of plasticity is used by biologists to refer to phenomena that explain developmental or physiological variation (chap. 6). A second approach concerns cases where the concept of plasticity is used to account for phenomena that are specifically subject to natural selection and are therefore understood from an evolutionary perspective (chap. 8). From there, I have considered how this distinction highlights how biologists can continue to think about the question of variation independently of evolution and how their thinking about plasticity (in the developmental sense) can be articulated within an evolutionary framework. The analysis of the phenomena of induction (chap. 6) and regulation (chap. 7) on the one hand and canalization and adaptation (chap. 8) on the other has allowed me to clarify the differences between the approach of analyzing variation for its own sake and the approach that consists in examining variation from an evolutionary perspective. In so doing I have been able to show how the concept of plasticity must be considered as a boundary concept.

The idea of a boundary, which underlines certain characteristics of the concept of plasticity, contrary to what one might think is never used to explain only a permanent phenomenon or to refer only to the idea of an integration of the fields of development and evolution but rather serves to constitute a singular space within which the biologist can place in a coherent way all that is in the nature of the signal, the ephemeral, the in-between. From this point of view, the concept of plasticity makes it possible to mark certain limits rather than sealing off the links between the different explanations available. The second important gain is that the reference to the idea of a boundary also makes it possible to think of the phenomenon of plasticity itself in a different, more illuminating way, not as a place of integration of different biological signals but as a place where these different signals still manifest themselves distinctly from one another.

My analysis of the concept of plasticity in the history of the life sciences in the first part of the book allows me to offer a new look at the deep stakes of the contemporary debate on phenotypic plasticity. I thus propose a detailed epistemological analysis of the concept of plasticity in contemporary biology

in the second part of the book. The debate, in which biologists have begun
to question how to qualify the importance of genetic information relative to
other factors (such as environmental or developmental factors) at work in
the generation of form, is part of a rapidly expanding field of research in contemporary biology: that of evo-devo. To understand how biologists who are
part of this trend think about the plasticity of living organisms, I do not limit
the analysis to theoretical work but also examine experimental approaches
carried out in this specific field (chap. 9). Now, with the expressed will of
some biologists and theorists to break with the "genocentric" tradition introduced by the modern synthesis, the question of plasticity has become the
center of attention, so much so that new definitions of the general concept of
plasticity have been proposed and continue to be hotly debated. I emphasize
that these definitions often retain parts, sometimes even the intention or the
general framework, of the definitions of the operative concept of phenotypic
plasticity. This has led me, in the concluding chapter of this book (chap. 10),
to suggest that the persistence of the concept of plasticity in the history of the
life sciences may be linked to the existence of certain epistemic paradigms.
By applying the idea of an epistemic paradigm to the concept of plasticity, I
highlight two features that may be essential for contemporary biology: (1)
the fact that reference to the concept of plasticity allows the biologist to insist
on certain characteristics of living beings that have not been observed (or
have not been taken into account) from both an empirical and a phenomenological point of view, and (2) the fact that the use of the concept is a means
for the biologist to pay attention to the possible blockages or obstacles that
the reference to the phenomenon of plasticity is likely to introduce in the
general reflection of the biologists and in particular of those who continue
to cultivate that immemorial ambition of achieving a unified and coherent
knowledge of the living world.

PART 1

Concepts of Plasticity in the History of the Life Sciences

Theories of Plasticity in the
Philosophy of Nature

It is a plea for us to listen, despite our fundamental differences of metaphysic and of
method, to some of the tenets that Aristotle, as a biologist-philosopher, advocated long
ago, and to try to interpret them in ways that could be useful to us as we attempt to
articulate and revise our conception of what the investigation and the knowledge of
living nature are.

MARJORIE GRENE (1974, 104)

While the term *plasticity* was first seen as a concept of metaphysics at the
time of Aristotle, it has now mostly become a scientific and an operational
concept of biology. Nevertheless, some of the old metaphysical associations
with the concept of plasticity still resonate when biologists use the term today. Following the advice of Marjorie Grene, I unravel the understanding of
the concept of plasticity not only by Aristotle but also by some later philosophers of nature. By tracing the intellectual history of the concept of plasticity,
I show which ideas continue to influence current conceptions of plasticity in
the life sciences.

Although some obvious uses of the term appear in the history of philosophy, such as the seventeenth-century reference to "the Plastick Life of Nature"
by the Cambridge Platonist philosophers Henry More and Ralph Cudworth,
the term has also occasionally disappeared, although the ideas associated
with the concept remain. This was the case, for example, with the work of
Christian von Wolff in the early eighteenth century. In a sense, these uses
and non-uses of plastic- terms[1] throughout the history of natural philosophy
could be seen as a feature of the concept of plasticity, since it largely disappeared when the concept of gene was invented even though it was sometimes
used by biologists who rejected genetic determinism. It also reveals the specific difficulties that contemporary biologists have encountered in defining
the term clearly and unambiguously when they use it. To understand and
analyze the use of plasticity in contemporary biology, it seems necessary to
return to its roots in natural philosophy.

This chapter has two objectives: to trace the genealogy of the concept of
plasticity in natural philosophy and to identify the philosophical (sometimes

metaphysical) ideas related to the concept of plasticity that still influence the use of the term in biology today.

Plasticity and the Generation of Animals within Natural Philosophy

As the historian Jane Oppenheimer has pointed out, Aristotle's work "consti-tute[d] a collection of knowledge on the development of the chick which be-came the foundation on which all embryology was to build" (Oppenheimer 1967, 120). In this context, Aristotle's use of plastic- terms to describe animal generation (a topic that included both its reproductive and developmen-tal aspects) still resonates throughout the history of the concept, including its modern usage. Therefore, an analysis of the work of the "philosopher-biologist" Aristotle seems to be one of the fundamental bases in our search for the origins of the concept of plasticity.

ARISTOTLE'S UNDERSTANDING OF GENERATION AND HIS USE OF *PLASTIKOS* AND *PLATTEIN*

As many commentators have already noted, Aristotle was definitely wrong about many of the biological phenomena that he attempted to describe, and on several occasions, he was also wrong about the interpretations he gave to his observations on the origin of generation and, in particular, the role he assigned to sperm. But he was not the only one in antiquity to be wrong on this issue (the pre-Socratics, such as Democritus and Empedo-cles, had also made false assumptions about the role of sperm). However, Aristotle was the first philosopher to use empirical and rational arguments to talk about generation as opposed to the mythical tales of a number of his predecessors. This enabled him to accomplish notable advances in the analysis of living beings.

For instance, by distinguishing the *telos* or the "idea"[2] from the "matter,"[3] Aristotle separated the "active" and "passive" components in the descrip-tion of the process of generation (Needham 1959, 43). The "active" compo-nent referred to what modified the developing entities, while the "passive" component referred to what was modified during the developmental process (Cooper 1988). This distinction formed the basis of his notion of plasticity. Aristotle referred to two main plastic- terms in his work: *plastikos* and *plat-tein*. While the Greek noun *plastikos* (the modeler or sculptor) was rarely used in Aristotle's work, the verb *plattein* (meaning to make or shape) ap-peared on many occasions.

In comparing the activity of nature to that of a craftsman demiurge, Aristotle elaborates on a metaphor common in antiquity (especially in the works of Plato), which allows him to develop a number of more technical meanings for the term *plattein* in relation to the generation of animals. For instance, when Aristotle compares the male seed to the activity of sculptors on ductile material (Solmsen 1963, 473–96), he uses the term *plattein* in an active sense, "to shape," and in a passive sense, "to be shaped," that is, the action of responding, of adapting to input from the active side, as one would say of clay. This distinction between a passive and an active component in the use of the term *plattein* echoes the dual understanding of generation as both an active (that which has altered developing entities) and a passive (that which is altered in the developmental process) process.

In his *Generation of Animals*, when Aristotle (1972) attempts to determine the role of the female in relation to that of the male in the formation of the embryo, the verb *plattein* is used in its active sense to designate the action of the males in "shaping" (*tois plattousin*) a material that may be related to clay (the female). Similarly, *On the Parts of Animals*, the term *plattein* appears again in its active sense when Aristotle attempts to explain the formation of flesh around the bone structure (Aristotle 2002, 2.9.654b27–32). If the metaphor of the modeler seems to give the active sense of the verb *plattein* some primacy, Aristotle also acknowledges the passive sense of the term when he describes the process of the body's formation, and the "middle-passive" sense appears with regard to the description of the generation of snakes (Aristotle 2002, 4.9.676b6–10).[4] In the latter case, however, the discussion is clearly about what is altered in the developmental process of the snakes.

From these few examples, it seems that Aristotle used the duality of the concept of plasticity to highlight either or both the active and the passive component in the process of generation. In a way, the reference to this term (*plattein, plasthenai*) allowed him to place form and matter in the same reality. This was something new compared to his predecessors, such as Plato, who had elevated form into another world above the world of material realities.

Aristotle, "by envisioning form as part of actuality rather than something above it, brought biological material to be directly investigable by the sense organs" (Oppenheimer 1967, 121). With Aristotle, the questions of the origin of the form and the organization of matter were brought together in a dynamic understanding of the process of generation. This conception of generation must be regarded as one of the major contributions of Aristotle to embryology (Needham 1959, 54–55).[5]

Moreover, by distinguishing the idea of preformation (the unfolding of preexisting structures) from that of epigenesis (the generation de novo),

Aristotle stated the possibility of a structure emerging from matter. The analogies between the process of generation and the artistic processes of net knitting, portrait painting, or statue modeling take on a new dimension. For Plato, the ideal mesh would already have been woven, the portrait completed, and the mold shaped. For Aristotle, even if the structure can be discussed separately—as can the order of the effective development or that of logical existence—a form cannot belong to a world other than that of matter: form is linked to matter and vice versa. From this point on, *both realities ("matter" and "form") appear in the same world, whereas for Plato they were "inhabitants" of different worlds.*

In this context, where Aristotle placed all beings in the material realm, plasticity is no longer just part of the world of matter with its active sense nor just part of the world of form with its passive sense; it is part of a dynamic world that includes both senses. Thanks to this new framework, Aristotle provided the first elements for understanding what would later be called not generation but *development*, a process involving both generation and formation.

However, by underlining the distinction between an active and a passive component of generation, Aristotle made it possible to consider epigenesis and preformation separately and then to link them intimately. It is thus with Aristotle that it becomes possible to think of both the "ability to shape" as well as the "ability to adopt a form"—also corresponding to the two active and passive meanings implicit in the philosophical concept of plasticity—within the same rational conception of the natural world.[6] These philosophical debates on generation, linked to the concept of plasticity, would resurface with the European Renaissance, revived by the humanism of the sixteenth century and the return of Aristotelian thought.

HARVEY'S *VIS PLASTICA* IN THE SETTLEMENT OF EPIGENESIS

William Harvey (1578–1657), whose discovery of blood circulation marked a major episode in the history of medicine (Harvey 1628), explores an active type of plasticity in his important work on generation (Harvey 1651). *Exercitationes de Generatione Animalium* remained engraved in the memories of some of his contemporaries because of the image on the frontispiece, which showed Jupiter opening a box in the shape of an egg with the inscription: "From the egg comes all" (Ex ovo omnia). Because of this image, Harvey was often credited with being the father of *ovism preformationism*—every organ would have been preformed in the ovum (Lawrence 2008). Despite this common view, Harvey mostly adhered to the Aristotelian theory of epigenesis—each organ is created de novo from the egg—because he thought this was the mode of generation of the most perfect animals (with red blood) (Harvey 1651, 334–35).

At the time, Harvey did not talk about "preformation" in contrast to epigenesis, as it was later in history (see Gould 1977, Roe [1981] 2003),[7] but he distinguished epigenesis from metamorphosis. Consequently, metamorphosis was not a theoretical alternative to epigenesis but rather a specific mode of generation that affects only some living beings. In any case, it was not the norm: it concerned those animals for which the material was already present in sufficient quantity, and the parts of the animal were then reconfigured from this material (this was the case with butterflies that develop from caterpillars). For Harvey, the process of epigenesis was similar to nutrition since the different parts were redistributed after ingestion and clumping (Harvey 1651, 339). This more elaborate definition of epigenesis (as opposed to Aristotle's somewhat allusive view) as a progressive establishment, which Harvey illustrated with numerous examples to show how body structures emerge from a central point to develop step-by-step, constitutes his major contribution to the history of embryology.

A close reading of Harvey's essay on generation reveals the multiple references he makes to the notion of *vis plastica* (translated from Latin into English by Robert Willis in Harvey [1847] as "plastic force" or "plastic power"). Harvey's reference to the *vis plastica* enabled him to detail and sharpen his understanding of the process of generation and its implications. With this notion he developed the idea that generation is an active and creative process that leads to the production of living beings. His focus was mainly on an implicit "active" meaning of plasticity. However, he understood the *vis plastica* as a placeholder for a complex web of relationships rather than as a particular "force":

> But as soon as the egg, under the influence of the gentle warmth of the incubating hen, or of warmth derived from another source, begins to pullulate, this spot forthwith dilates, and expands like the pupil of the eye, and from thence, as the grand centre of the egg, the latent *plastic force* breaks forth and germinates. This first commencement of the chick, however, so far as I am aware, has not yet been observed by any one. (Harvey 1651, 229, my emphasis)

Harvey described the emergence of the plastic force (*vis plastica*) and showed that the phenomenon required a sort of curing process, a latency period, in order to play its role in the further development of the embryo. The notion of *vis plastica* is thus inevitably bound to an Aristotelian epigenetic view of generation, achieved "per degree," "bit by bit," as the plastic force "breaks forth and germinates" rather than preexists. Somehow Harvey would associate the "active" meaning of the concept of plasticity, through the notion of *vis plastica*, with the process of generation. He thought that whether one accepted or rejected the idea of a vital principle (or of a soul) inherent in the egg, one had to refer to a principle whose influence made possible the continuity of the circle from egg to chick and from chick to egg (Harvey 1651, 285–86). Harvey equated this principle with a divine and eternal principle, a superior intelligence that manifested itself during generation through the *vis plastica*.

Some aspects of the Aristotelian dialectic between form and matter appear here in Harvey's idea, since Harvey referred to the *vis plastica* as an intermediary between form and matter. Although the "active" meaning of plasticity was emphasized in the "plastic force," the "passive" meaning, understood as "susceptibility to adopt a form," was also emphasized, as this quotation shows:

> The processes of formation and growth are simultaneous. In the former the plastic force cuts up, and distributes, and reduces into limbs the same homogeneous material; and makes out of homogeneous material organs which are dissimilar. But in the latter, while it creates in succession parts which are differently and variously distributed, it requires and makes a material which is also various in its nature, and variously distributed, and such as is now adapted to the formation of one part, now of another; on which account we believe the perfect hen's-egg to be constituted of various parts. (Harvey 1651, 320)

Harvey first highlights the action of the plastic force that shaped the matter into limbs and organs. It is what he calls "formation," but this is quite different from Aristotle's formation (the ability to adopt a form), since it refers more to shaping. Harvey then refers to the plastic force as something that "requires" a material of a different nature and distribution. In other words, Harvey explains that "plasticity," to be expressed as a "potentiality to shape" (in an active manner), requires a variability in the distribution of the form (plasticity in a passive manner). For Harvey, plasticity, with *vis plastica*, becomes an intermediary between the inert and homogenous material *and* the specific and determined forms.

For Harvey, unlike Aristotle, it was not possible to think of the production of an egg without the equal concomitant contribution of the female and the male. Aristotle believed that the male was the most important factor in the egg formation. Harvey also eliminated Aristotle's division between passive matter (the outcome of the female) and active form (the outcome of the male). He emphasized the materiality of form and showed that form depends as much on the male as on the female. For this reason, Harvey continued to ascribe some materiality to the *vis plastica* and even attempted to locate it, as it appears in the following quotation: "But in the seeds of all plants there is a gemmule or bud of such a kind, so small that if the top only, a very point, be lost, all hope of propagation is immediately destroyed; in so small a particle does all the plastic power of the future tree seem lodged!" (Harvey 1651, 320–21). For Harvey, all the plastic potential of the tree seems to be located at the end of the gemmule or bud of the future tree. In locating the plastic force, Harvey also ascribed some materiality to the faculty.

Aristotle's and Harvey's use of plastic- terms (*plastikos* and *plattein* for Aristotle, *vis plastica* for Harvey) appears mainly in the context of both developing ideas about the generation of living animals. Whether their use of plastic- terms was associated with an active (what modifies the developing entities) or a passive (what is modified during the developmental process) conception of generation, it quickly translated for later commentators into the idea that they were both in favor of an epigenesist view of generation (the development of organs happening gradually, part by part), soon to be opposed to a "preformationist" view of generation (where all the body organs are already present in the embryo). The concept of preformation does not appear as a theoretical opposition to epigenesis until the late eighteenth century (see Roe 1979; Roe [1981] 2003; Maienschein 2003).

With Harvey, the observation and description of the process of generation became more systematic, and some of the theoretical ideas developed by Aristotle could no longer be supported by observations. For Harvey, generation seemed to depend on the action of a "plastic force," a kind of efficient cause that acted as an intermediary between matter and form. This intermediary, necessary to describe and explain generation, was depicted neither as a nonmaterial force nor as a material element that could be localized. The localization of the plastic force remained without a definitive answer for Harvey (even though he seemed to be in favor of a materiality). It was not until the seventeenth century that the question was reconsidered in light of mechanistic conceptions of living beings. In any case, from that moment on, the active sense of plasticity, understood as the "ability to shape," was the one that was adopted by the philosophers of nature of the seventeenth century.

The Theories of Plastic Nature(s) in Reaction against Cartesianism

INITIAL CONTEXT: THE SEVENTEENTH CENTURY'S MECHANICAL PHILOSOPHY

A year after Harvey's death in 1657, a new era began in the history of natural philosophy and of theories of generation. The complete work of the French mathematician, philosopher, theologian, and astronomer Pierre Gassendi

(1592–1655) was published (Gassendi 1658). His treatise *De Generatione Ani-malium* (vol. 2, bk 4 of his *Opera Omnia*), alongside philosopher René Descartes's (1596–1650) treatise *La Formation du Fœtus* (Descartes [1677] 2018), were among the most influential texts on theories of generation in this era. As the biologist and historian Joseph Needham showed in his *History of Embryology* "Descartes and Gassendi built up an embryology more *geometrico demonstrata*, in which the facts where relegated to an inferior position and the theory was all" (Needham [1959] 2015, 235; see also Duchesneau 1998). Indeed, in his treatise, Descartes sought to explain living phenomena through the same laws as those that govern inanimate matter. Since he mainly relied on mechanical principles to explain the development of the embryo, he defended a conception of generation quite different from Aristotle or Harvey. Descartes's position involved a methodological subordination of organic phenomena to various models of the animal-machine. His explanations of vital functions, such as those of generation, were hypothetico-deductive constructions based on mechanical analogies (Descartes [1649] 2017). This led Descartes notably to consider animals as automata in their behavior and to consider animals to be organic machines.

At the time he was among the very first theoreticians to attempt to think about living beings without any reference to powers, forces, or principles other than those of mechanical or material causality. For Descartes, material causality could explain generation on its own. The radical nature of the Cartesian model would have a severe impact in the life sciences. Even if his views about generation were soon criticized and even rejected by some of his successors, from then on, the mechanistic views have not ceased to have followers in the history of the life sciences.

In the seventeenth century, the Cambridge Platonists were, in some way, the first characters of a long and multiform tradition that rejected the Cartesian "mechanistic" conception of generation.[8] For those Platonists, the limit of the Cartesian model was not its subordination to a mechanical causality in the explanation of generation but the fact that in this model a principle governing, adjusting, and combining all the different mechanisms together was lacking. The Cambridge Platonists thought that such a principle improved the explanations of the phenomena of generation by not restricting generation to the solely physico-mechanistic processes.

More precisely, two of these Cambridge Platonists, Henry More (1614–1687) in *The Immortality of the Soul* ([1659] 1987) and Ralph Cudworth (1617–1688) in *The True Intellectual System of the Universe* ([1678] 1829) took radical positions against the Cartesian model about generation and in particular against the absence of any explanatory principle by which the different mechanical processes of generation would be implemented. For both More and Cudworth, a soul or an intellect was the principle that provided direction for the mechanical processes. In other words they claimed that such principles were necessary for "the production of harmonious, functional and integrated organizations that constitute living beings" (Duchesneau 1998, 14). It is in this context that they first referred, in the seventeenth century, to the notion of "plastic nature."

Already in 1950 historian William B. Hunter argued that the Cambridge Platonists's notion of "plastic nature" established a "convenient catch-all for almost every sub-sensitive manifestation of life." In other words, Hunter argued that for these authors the term was used to represent "not merely matter in motion but some manifestation of spirit of incorporeal substance," which formed matter into various organisms (Hunter 1950, 200–201).

The plastic nature concept did not simply represent material in movement but rather was mostly the expression of a spirit or of an immaterial substance. For those who subscribe to it, the shaping of the organic matter seemed to rely on a kind of architectonic force necessary to biological organization, a force that they called plastic nature. Through their metaphysical speculations, More and Cudworth helped legitimize thinking of living beings differently from the mainstream Cartesian mechanistic view.

I expand on Hunter's position and argue that from the time of the Cambridge Platonists, the formulation of the notion of plastic nature provided a benchmark in the history of the concept of plasticity. From this period, plastic-terms would often appear in clear connection with the life sciences without being limited to generation only, as it was in Aristotle's or Harvey's writings.

MORE'S PLASTIC POWER OF THE SOUL AS AN ORGANIZING PRINCIPLE

More, like Cudworth, was a Cambridge Platonist, that is to say he rejected Aristotle's physics. However, some of the Cambridge Platonist's ideas concerning the relationship of the soul and matter were a reminder of Aristotle's hylomorphism—the connection between the soul and the body. But like Plato and contrary to Aristotle, the Cambridge Platonists thought that the soul took priority in the matter.[9]

Contrary to Descartes, Henry More did not stress an independence between the material and the immaterial realms. On the contrary, More was rather looking for the principles that could help account for the soul and matter together. Understanding the integrated production of the living beings did not require, for him, identifying how the soul and matter mechanically joined together but rather required assuming that both the soul and matter each included a formative part. For More, these two parts were to be thought of as independent and, tellingly, not due to any influence of a demiurge.

More had a personal interpretation of the Aristotelian thesis as it had been revised by Harvey. More believed in a gradation of the soul from the "*plastic faculty*"—the ability to take on functional material form—to the perceptive, imaginative, intellective and other faculties. In a way, the soul became more material in these faculties. He then condensed this gradation into a duality of essential faculties of the Soul: perception was on one side, and organization was on the other. While perception depended on the activity of the *sensorium commune* of the Soul (More [1677] 1987, 2.4.2, 2.6.1, and 2.6.9; 3.2.4), organization (or the capacity of formation) depended on the "plastick power of the Soule," to which were connected the vital functions, such as the beating of the heart, breathing, blood circulation, tonicity, and muscle contraction (More [1677] 1987, 2.8.6). More rejected Descartes's theory according to which the vital functions of the body could be described as for a machine (Duchesneau 1998, 103). He thought that the soul was bound to the whole body and, thus, present in all the parts of the body while being indivisible.

In contrast, for Descartes, the "Animal Spirits"[10] were located in the pineal gland and were responsible for voluntary movement (Descartes [1641] 2016, [1649] 2017). More was opposed to Descartes's interpretation: he rejected the idea according to which matter was capable of forming living bodies only through its faculty and organization. More proposed the idea of an intellectual nature of causality that was opposed to a simple mechanical causality. This intellectual nature of causality was manifested through the plastic faculty, which itself constituted and integrator principle between soul and matter in the biological organizations.

But before even trying to define living beings, one of More's purposes was to denunciate the mechanistic approach for its inability to account for the specificity of living phenomena. More rejected the mechanistic model from an epistemological point of view but not from an ontological point of view, as some vitalists would later do. Indeed, while the vitalists rejected the idea that all events were identical to mechanical phenomena, More rejected the idea that the laws of mechanical phenomena were sufficient to explain living phenomena (see Huneman and Rey 2007, 221). In other words, More was not an essential vitalist even though he endorsed a kind of explanatory vitalism. Therefore, More was not a vitalist per se, despite his use of the concept of soul, because contrary to the vitalist doctrine, his purpose was not to define the specificity of the living beings but rather to understand the organization and generation of living beings. For both More and Cudworth, Descartes had failed to offer a proper account of the living processes in this regard because of his omission of an integrating principle such as soul.

This critique was quite significant because many further attempts by natural philosophers to characterize generation would involve making reference to integrative principles. From then on, the notion of "plastic nature" also became a spearhead of opposition to mechanistic views.

CUDWORTH'S PLASTIC NATURE: A MENTAL CAUSALITY EXPLAINING ORGANIC PHENOMENA

In *The True Intellectual System of the Universe*, Ralph Cudworth ([1678] 1829) tried to overcome a twofold theological difficulty concerning the productions of nature: either God was directly making everything, and this exclusive activity was inconsistent with God's dignity, or God was an idle spectator of a world operating with its own rules, and this idleness contradicted God's perfect nature. The reference to a plastic principle allowed Cudworth to escape this dilemma: God became active but indirectly, through plastic nature, which was the relay of God's action (and also the testimony of His existence).

Cudworth thought that "all things [were derived] from a mere fortuitous principle, from nature and chance; that is, the unguided motion of matter, without any plastic artificialness or methodicalness, either in the whole universe, or the parts of it" (Cudworth [1678] 1829, 245). In other words, Cudworth thought that nature acted through generation in conformity with a design that it could not conceive, receiving it from the architectonic wisdom of God but executing it through its own causality without being linked to the artifacts of the intellect that one could imagine connected to the soul. In a way, his position was faithful to William Harvey, who had attempted before him to give an account of this causal process.

For Cudworth the plastic nature corresponded to a determined and specific causality, which he qualified in terms of "mental causality" (Cudworth 1820, 331). This argument of Cudworth is really where the ontological opposition to the Cartesian model lies: the "mental causality," meaning a final causality ("disposing things in order to ends" Cudworth 1820, 331), enabled him to give an account of organic phenomena that were not limited to matter or any kind of motion of the matter while explaining nature's order and regularity. Thanks to this mental causality, Cudworth argued that the plastic nature should be conceived as responsible for the framing and the conservation of animals' bodies.

Cudworth argument was that this plastic nature was not only in animals' bodies but everywhere in the universe, explaining its general order and disposition. For Cudworth, "Nature [was] either a lower power, or faculty of some conscious soul, or else an inferior kind of life by itself, depending upon a superior soul" (Cudworth 1820, 356). The plastic nature is thus, for Cudworth, a finalized and finalizing agent immanent to the general material order, embodied in the animal microcosms in the form of psychic principles, and responsible for the morphogenetic functions of sensitivity and locomotion. In Cudworth's view, the plastic nature acts immediately on matter as an inward principle. What can we learn from More's and Cudworth's visions of plastic nature?

Beyond Animal Generation: The Plastic Nature Explains Other Living Products

The notion of plastic nature was first built in a rational manner on opposition to Cartesian mechanism. However, the idea of an opposition to mechanism, whatever form it took, would later bring the concept of plasticity together with occult vital principles, such as those that appeared in the vitalist doctrine (traditionally opposed to the mechanist approach). The reason is probably embedded in the confusion between ontological vitalism and epistemological vitalism. Contemporaries of the Cambridge Platonists took their view as ontological vitalism in which they were mostly referring to an epistemological or explanatory vitalism.

Even if the Cambridge Platonists' view was mistakenly lumped in with those of the ontological vitalists, the intuitive correlation between the use of the term *plastic* and the vitalists' doctrine would continue to persist. This can partly be explained by the Cambridge Platonists' lack of influence on their contemporaries concerning generation and compared to the great success of the mechanism positions in natural philosophy and then in the life sciences.

However, in natural philosophy, the notion of plastic nature became, thanks to the Cambridge Platonists, intimately linked to the idea of a property of soul, of an intellective faculty (the plastic power of the soul for More and the "mental causality" for Cudworth). This idea, linked to plastic- terms in philosophy, came to complete the Aristotelian view in the explanation of the generation of living beings. The introduction of an intellective faculty broadened the scope of application of the concept of plasticity since it was not anymore only linked to processes of organogenesis but also referred to all the other productions of living beings, including the animals' behaviors as much as their conservation.

While More and Cudworth's influence would remain quite underground after they proposed their model, their approach somehow reemerged after the first Newtonian success and the strengthening of empirical and phenomenal methodologies in the natural sciences (Duchesneau 1998, 181). Indeed, the Cambridge Platonists' approach established the basis for a view of nature in which the concept of plasticity contributed to the explanation of the unique organization and integrity of living bodies. But if the reemergence of the notion of plastic nature in natural philosophy is not incidental (since Leibniz drew on More and Cudworth), its resurgence in Leibniz's work is combined with the transition from the singular *plastic nature* to the plural *plastic natures* (natures plastiques in French) (Leibniz [1705] 1840) and with access to a wider audience.

LEIBNIZ'S MATERIAL PLASTIC NATURES: ORGANIZING PRINCIPLES WITHIN LIVING BODIES

The Platonists' critique of the Cartesian model was precisely about the inability of the mechanistic model to offer a formal principle that could explain the harmonious and integrated organization of living beings. Leibniz was able to bypass this critique because of the formal causal role of entelechy in his inclusive model, which, in a way, took into account both the Platonists and the mechanist approach: "Nature's machines—living bodies, that is—are machines even in their smallest parts, right down to infinity. That is what makes the difference between nature and artifice, that is, between divine artifice and our artifice" (Leibniz [1714] 2014, §64). After Leibniz's reference to a formal causality in his model, he was free to focus on mechanistic explanations, and the influence of Cartesian mechanism appeared throughout Leibniz's use of the notion of "nature's machines."

However, Leibniz's articulation of the plastic natures of nature's machines would only appear later, in his *Considération sur les principes de vie et sur les natures plastiques* ([1705] 1840; Considerations on vital principles and plastic natures). In this essay, Leibniz described organic bodies as "material plastic natures." This notion illustrated the articulation between Leibniz's unifying principle—the entelechy–that he used to explain the functional order of nature along with the mechanical connections between the parts on which the integrated structure relied (Leibniz [1705] 1840).

The conciliation between the theory of plastic natures and the mechanistic approach became possible in Leibniz's integrated and systemic view because of his "material" understanding of the plastic natures. Leibniz's reference, for the first time, to the notion of "organism" illustrates his conciliation between empiricism and mechanism but also the fact that he distinguished himself from other more radical Cartesian approaches (e.g., those of Herman

Boerhaave's chemical and mathematical decompositions). Furthermore, the notion of "organism" allowed him to articulate both bodily and psychic agencies.

While the differences between the Cambridge Platonists' and Leibniz's use of the plastic nature concept might seem, at first, minor, they in fact constitute a major metaphysical turn in the understanding of living organizations.

Indeed, while the Cambridge Platonists had remained faithful to a kind of "holistic" view concerning living beings, a view itself based on the conception of generation of animals from antiquity (i.e., the generation process was confined to the animal in its entirety, which was viewed as a unified whole), Leibniz, for his part, promoted a mereological view of organisms and brought the plastic faculty to the sub-elements of organisms. This distinction between a holistic view of organization and generation and a mereological view of organization would be perpetuated in all the latter understandings of plasticity in the life sciences, wherein plasticity may be seen as the property of the organism as a whole, or a property of sub- or supraclasses such as cells or species.

On the one hand, in Aristotle writings the concept of plasticity appeared mostly in relation with the theory of generation later mostly understood as epigenesis (the progressive development of the animal); it is at least from that time that the term *plastic* became synonymous with the "ability to generate" and that it was seen as a property of the whole animal (i.e., in a holistic view). On the other hand, with Leibniz's material plastic natures of organisms and despite anti-Cartesian Platonists' use of the "plastic nature" concept, the concept of plasticity became compatible with a mechanistic view, which itself relied on a mereological and potentially reductionist view of living beings. In other words, even if Leibniz adopted a mereological view of living beings, the plastic natures expressed at the level of simple entities were used to explain organic bodies' complexity. However, the transition from the Cambridge Platonists to Leibniz, from the singular of "plastic nature" to the plural "plastic natures," illustrates the transition from plasticity understood as a holistic or general property to plasticity understood as a systemic property of a composed whole, or of the simplest entities of living organisms.[11] While this work has contributed greatly to the development of a general view of plasticity in life, the more precise connection of the concept of plasticity with the life sciences has much to do with the thinking of another author who has not, however, mentioned the notion of plasticity in one form or another in his work. This is what I now examine.

From Metaphysics of Plasticity in Life
to Plasticity in the Life Sciences

THE ROLE OF CHRISTIAN VON WOLFF'S
METAPHYSICS IN THIS TRANSITION

Although the issue concerning the influence of Leibniz on Christian von Wolff's philosophy is still discussed among commentators, the similarity of their metaphysical approaches on nature seems obvious (Hettche and Dyck 2019; Corr 1975; Lenders 1971; École 1964). In both cases, the world and living bodies consisted of the interconnection of the smallest entities. The modes of interaction between these entities were based on space, time, or causal relations. These entities possessed inertia, an active force, and a motion force. Finally, their whole doctrines relied on a theory of simple substances or "elements," themselves being the ultimate compositional units of larger beings.

While it remains quite difficult to see precisely what the conceptual differences between the two philosophies were (see École 1964, 1979),[12] there is one obvious difference, which concerns their reference to the concept of plasticity: Leibniz referred to the notion of plastic natures, whereas no mention is made to this controversial doctrine in von Wolff's work. However, despite this difference, it remains interesting to look at the way von Wolff addressed the same metaphysical issues that Leibniz tried to enlighten when he referred to this notion.

As claimed previously, Leibniz adopted a mereological view of living beings wherein the plastic natures of the simplest entities explain the complexity of whole organic bodies. Both Leibniz and von Wolff formulated a critique of the Cartesian approach even though they claimed some continuity with it (Corr 1975, 253; Corr 1972).[13] While Leibniz went beyond the Cartesian model with his entelechy, his view of "material plastic natures" also promoted a renewed form of mechanism. Indeed, the system he proposed did not come into direct conflict with Descartes's mode of analysis of nature. Contrary to Leibniz, however, von Wolff understood the term *system* as including both an opinion and a hypothesis or a mode of scientific reasoning, themselves in relation to a precise doctrinal set (what he called "scientific thought" [Denkungsart Wissenschaftliche]) and with an organic or architectonic structure of disciplines (what he called the "doctrine" [Lehrgebäude]) (Corr 1972; Lenders 1971). In this regard von Wolff can be seen as an epistemological constructivist. Von Wolff agreed with Leibniz's understanding of the "system" in Leibniz's "system of a preestablished harmony"—the system is what gathers the soul and the body together—but he thought that Leibniz had failed to offer a broader definition of the system since he did not give any kind of methodology in order to do so. Furthermore, von Wolff thought that Leibniz's model had failed to really offer a proper "integration" within the system in the way suggested by the Platonists throughout their critiques of the Cartesian model.

Therefore, while von Wolff adhered to the preestablished harmony theory, he thought that it was not per se a founding principle of the system. He thought that it was rather a characterization of the harmony between the soul and the body. For von Wolff, the role of the philosopher was not to continuously question the relationship between the soul and the body but to offer a justification of this relationship through a philosophical hypothesis. In other words, von Wolff was more interested in the change that could happen in the relationship between the soul and the body through experience than by the relationship itself. He thought that such a change was the result of an "influx" that was, in the sense of the physicists, an "action" that could cause material changes (von Wolff 1736).

While von Wolff did not agree in the existence of vital forces within simple substances (e.g., as claimed by Cambridge Platonists), he did propose, contrary to Descartes, a certain active force in the bodies based on the changes that were intrinsically happening to them because of the articulation of their components. For von Wolff, force relied on the interconnection between the simplest substances (Burns 1966). This definition of the force inherent in the bodies is, in a way, reminiscent of Harvey's "plastic force," which was defined as "some principle influencing this revolution from the fowl to the egg and from the egg back to the fowl, which gives them perpetuity" (Harvey 1651, 285–86). Von Wolff mainly described this force as "a constant need to act," but he also thought that it could be described as any other physical force. Leibniz, on his side, decomposed the force within the bodies into two parts: the "primary force" that he also called the "dead force" (vis mortua) with no motion involved (e.g., the force of a stone in a sling when it is still in it), and the ordinary force linked to a clear motion that Leibniz called the "active force" (vis viva).

Von Wolff, on his side, did not consider that the active force of the bodies was a real substance, but that it was rather the effect of the active force proper on the simple elements of the bodies in their interconnection. Consequently, even though von Wolff did not mention plastic natures, his understanding of the interconnection between the primary entities and their physical forces was in line with the Platonists' view. But thanks to his empirical approach, von Wolff's view may be distinguished from theirs along with those of Leibniz because he adopted a holistic and fundamentally systemic view of organization irreconcilable with the mereological view of Leibniz. Leibniz's view would then be taken over by supporters of the reductionist view of organized bodies. While it remained difficult for von Wolff to articulate his metaphysical view with the physical order of things, he offered a rationalist view concerning the structure of living bodies that would have a noticeable influence on the emerging field of embryology, as I will show in the next chapter.

WOLFF'S RATIONALIST METAPHYSICS OF BEING

Where Leibniz focused on the entities and parts, von Wolff insisted on the connection and articulation of these elements together. For von Wolff, the changes happening in the bodies were to be explained because of the changes happening in the organization of the system and not because of the nature of the entities. Because every entity produced a constant and continuous series of changes, time was for von Wolff the objective measure of these changes. Understanding the role of time in von Wolff's philosophy shows how he saw the interaction between entities. Furthermore, it is this conception of time that would later highly influence the first empirical reflections of Caspar Friedrich Wolff (one famous student of Christian von Wolff) on embryology (see chap. 2).

For Christian von Wolff, entities' interaction was first and foremost temporal and not spatial. The series of internal changes to which the entity was subjected was the result of its own driving force as much as the driving forces of the other entities to which it was connected. For von Wolff, the forces resulting from the interaction of multiple simple entities give the idea one has of an extended object. Consequently, the change happening within the body (that one would call, in modern terms, "its plasticity") relied more on the interactions and forces to which it was submitted along its lifetime than on its initial nature. For von Wolff, this metaphysics of being, throughout the study of physical bodies, relied on a dynamical conception of change within the structure that did not rely only on "forces" or on the nature of parts themselves but on the physical interactions of these parts along the lifetime of the being in question.

Let us now summarize von Wolff's position in the debate on the plastic natures. Von Wolff adopted a view on the structure of being and the interconnection of its parts different from that of Leibniz in many respects. However, this view, concerning the articulation of primary entities and their physical forces, was clearly embedded in the discussions about the plastic natures and their role on the entities of nature. These discussions had been initiated by the Cambridge Platonists and then taken up by Leibniz. While von Wolff did not mention at any time in his work the notion of "plastic nature" or the idea of a "plastic force," he continued the reflections of his predecessors concerning the structure and the organization of being, sometimes even on a more empirical level (compared to some of his metaphysicians' predecessors).

Consequently, von Wolff promoted a dynamical understanding of change (and also of plasticity) that would later characterize the description of the structure of every being. However, von Wolff did not limit his inquiry to organic bodies and tried to give a rational explanation for all physical bodies. By continuing Leibniz's work, von Wolff unwittingly took a position in the discussions concerning plastic natures (which he implied in his understanding of the organization of organic bodies). Von Wolff's position concerning complex bodies led him to adopt an empirical and phenomenal view of nature. In his opinion complex bodies relied mostly on the internal interactions that connected and disconnected them even more than on their constitutive primary entities. Such a dynamical approach, which was not present in Leibniz's view because he mostly favored a structural and mereological view, would later lead other thinkers to connect the issue of plasticity in life science with the issue of embryonic development. Indeed, von Wolff's discussions on the structure of being would have an impact on the surgeon, anatomist, and embryologist Caspar Wolff, who took over his philosophy teacher Christian von Wolff's terminology in some of his discussions concerning how the embryo is structurally organized.[14]

Consequently, while Leibniz, had offered the possibility to think about plasticity as an emergent property from the parts of a whole, von Wolff would insist on the dynamical characteristic of a system. When transferred to living beings, such a view highlights the possibility of thinking about plasticity as a dispositional property—a definition that will be then largely developed in embryology.

Philosophical Assumptions Concerning Plasticity in the Study of Living Beings

Through this analysis of the genealogy of the concept of plasticity in natural philosophy, I have shown that the idea of plasticity in the life sciences was first raised in relation to two issues: the generation of living beings and the origin of their shape and structure. These two initial issues are today the two possible ways of understanding plasticity in biology, depending on whether one understands it in its "active" meaning as a "capacity to shape" or in its passive meaning as a "capacity to adopt different shapes." First, with Aristotle in antiquity, and then with Harvey in the period of the European Renaissance, it

was mainly the active meaning of the term, and thus the issue of the generation of living beings was at the center of discussion. This "active" plasticity is
the result of a "plastic force" acting within the developing bodies. From this
point on, the idea of plasticity in life was associated with the "epigenetic" view
of generation that would later be opposed to a "preformationist" view.

At the beginning of the seventeenth century, the mechanist theory of generation proposed by René Descartes would introduce a new theoretical context in which it was no longer only about the ontology of generation (what
it was) but also about the epistemology of generation (how to explain it).
It is this new theoretical context, as much as the debate concerning "plastic
nature," that would give birth to embryology as a scientific field in the eighteenth century.

The Cambridge Platonists rejected Cartesian mechanism in the explanation of generation and promoted the need to refer to a mental property—the
plastic nature—that was both an internal causal principle of matter and a
principle of a mental causality. For them, there was no clear opposition between the material and the immaterial; the issue was not about the definition
of life but about how to account for living phenomena in general. For this
reason, they adopted a holistic view on the organization within living beings
and were not limited in their reflections to the issue of generation. Leibniz's
view differed from that of the Cambridge Platonists in that he associated the
plastic natures of organic entities with their substantial form. Leibniz conceived the organization of living beings (of the organisms) in accordance
with Descartes's model: the explanation of the appearance and organization
of the form relies on the parts themselves. Leibniz adopted a mereological
and potentially "reductionist" view of living beings that relied mainly on their
primary entities. Christian von Wolff, a pupil of Leibniz, however, adopted a
different view since he thought that if there was an interconnection between
the entities at the origin of the bodies, the organization of these systems was
explained by the relations between the parts and not sufficiently by the qualities of the parts themselves. In other words, the organization of the bodies
did not rely on their substances but on the dynamical dispositions of their
entities.

This analysis of the origin of the notions of "plastic force" on the one side
and of "plastic nature(s)" on the other side has shown that some recurring
metaphysical issues linked with the concept of plasticity have persisted in the
history of natural philosophy, for instance, the understanding of the active
and passive part in the description of the form as much as the distinction of
the inert world from the living world or even the dynamical understanding of
the integration of the parts in the organization process. These issues underpin

an implicit and intuitive content of current use of the plastic- terms (e.g., "plasticity," "plastic ability," etc.) in biology, whether they refer to reflections about explanations of living beings' generation, their organization, or simply their forms and structures. The concept of plasticity continues to be used more or less explicitly in the life sciences today despite the emergence of more operational definitions of the term in specific subfields of biology (e.g., ecology, developmental biology, or genetics).

This becomes evident as in the course of describing the appearance of some of these definitions, some noticeable variations emerged between the different uses of the plastic- terms, as I will now show in the three following chapters (first in embryology and developmental biology and then in genetics and evolutionary biology). For this reason, the identification of some metaphysical issues in relation to the concept of plasticity in natural philosophy sheds new light on our current understanding of its contemporary uses. By taking a retrospective approach to the concept of plasticity, we not only describe the development of past views and past uses of the term but also accept interpreting them "in ways that could be useful to us as we attempt to articulate and revise our [current] conception" (Grene 1974, 104) of the concept.

2

The Plastic Embryo

[We can see] epigenesis and preformation [as] an antithesis which Aristotle was the
first to perceive, and the subsequent history of which is almost synonymous with the
history of embryology.

JOSEPH NEEDHAM (1959, 40).

Embryology is the branch of the life sciences devoted to the description of
morphological transformations of the fertilized egg into a complex organism
and the study of their causes. This field, which studies the embryo—and takes
its name from the Greek term *embruon*: *em* (into) and *bruein* (to swell, to
grow)—explains both the origin and the mechanisms of the internal growth
of the developing organism. Two major periods characterize the history
of embryology: descriptive embryology in the eighteenth century, with its
opposing rival theories of preformation and epigenesis, and experimental
embryology, which started in the second half of the nineteenth century and
became from the 1960s on so-called developmental biology because of the
rise to prominence of molecular biology, which ushered in a new conception
of embryology that also includes the study of postembryonic development
(the fetus and the newborn until the adult stage).

In 1959 Joseph Needham stated that understanding the links and implications of the epigenesis-preformation debate meant digging into the theoretical history of the field of embryology. Therefore, if the concept of plasticity
plays a key role in this debate, as I argue based on the philosophical considerations drawn in the previous chapter, it probably also has a decisive stake in
the theoretical history and development of embryology.

This long and complex history, its different theoretical and methodological turns linked to the broader history of ontogenesis theories, has indeed its
roots in some of the philosophical schools of thought discussed in the previous chapter. To demonstrate the continuity between the philosophical school
of thought and embryological works, this chapter explores how the concept
of plasticity was used in embryology over the course of this history until the
mid-twentieth century. By examining the use of plastic- terms or other terms

denoting similar ideas without the specific use of plastic- terms, we can determine how the history of ideas about plasticity has influenced this field over time. Because plastic- terms continued to be used with a broad philosophical meaning in the life sciences, more specialized uses also appeared in embryology. I not only look at the explicit use of plastic- terms in embryology through its textbooks and historiography but also identify "the strategic nonuse of implicit plasticity theories" in places where plastic concepts are present but other terms are used. This preliminary perspective is necessary to better understand the contemporary uses of plastic- terms in developmental biology later on (examined in chap. 7).

While embryology houses different methodological approaches, and these probably correspond to the main "epistemological episodes" of the field, each of these episodes continues in the sense that they are discussed and reintegrated over time rather than replaced by an entirely new paradigm. Even concepts like epigenesis and preformation, whose premises were defined in philosophy from antiquity and in a scientific context in the seventeenth century, have continued to play a part in the understanding of embryological phenomena. As discussed in the previous chapter, the seventeenth-century thinkers More and Cudworth used the idea of a "plastic nature" to talk about ontogenesis—the genesis of the being from a metaphysical point of view. After frequent use of the term *plastic* (and more precisely *plastic natures*) in the seventeenth century, the term tends to disappear in the eighteenth century before reappearing in the nineteenth century.

In the eighteenth century, study of generation became empirical with the emerging field of embryology. Living beings, by their ability to "develop," that is, to produce something new from old (previous generation) residues, could only be explained with the same laws of physics that explain the mechanical phenomena of the world. Caspar Friedrich Wolff (1734–1794), regarded as one of the "modern" founders of the field, initiated a new interest in epigenesis when preformationist views had become dominant. In Caspar Wolff's works the term *plastic* does not appear, as in the work of his great master Christian von Wolff, where the term was not to be found, and contrary to Leibniz, with whom he was often compared. The reason of such a non-use was probably for both Wolffs to avoid all vitalist and nonempirical connotations linked to the term. Indeed, at this period, the concept of plasticity had been frequently linked to animist, vitalist, and nonempirical views. However, Caspar Wolff, like Christian von Wolff before him, certainly contributed to laying the foundations of modern conceptions of plasticity in developmental biology. He was the first embryologist to formulate a general theoretical explanation on the dynamic organization of developing matter with detailed observations on embryonic development.

I first discuss what I have called the "implicit plasticity concepts" and look at ideas and theories developed in early descriptive embryology. I show that both active and passive definitions of plasticity appear when implicit plastic concepts are at stake. I examine how in parallel with the first development of embryology, a philosophical proposition was elaborated by Kant and Blumenbach that led to a prevalence of the passive sense of the term *plasticity* over its active sense. Then I examine the "revelation" of the first implicit concepts in the explicit uses of the adjective *plastic* in the embryology of the early nineteenth century. I look at its uses in different major theories (from tissues, layers to cells). Finally, I look at the shift of interest from plasticity used to characterize embryogenesis phenomena to plasticity used as an operational concept and the investigation of its causes through the study of inductive phenomena.

Implicit Plasticity Concepts in Early Embryology

THE REVIVAL OF EPIGENESIS IN THE EIGHTEENTH CENTURY

In 1672 the Italian biologist and physician Marcello Malpighi (1628–1694) had published one of the first empirical studies describing the development of a chick embryo observed through a microscope—*De Formatione Pulli in Ovo* (Malpighi 1673). The microscope brought a new methodological approach to the study of development, but it also promoted the basis for a new debate. The debate was between the epigenesists, who thought that embryos were formed de novo at each generation, and the preformationists, who thought that the organs were already present but very small in the ovum or the sperm cell. In the seventeenth and eighteenth century, in accordance with Malpighi's observations, preformationism became the predominant theory. It was thought that the developing animal was present in the egg, even though it seemed small, transparent and invisible at first. During development, the miniature animal grows, and its tissues become denser and therefore more visible.

The preformationist view was presented through different versions throughout both the seventeenth and eighteenth centuries. For instance, in the seventeenth century some scientists, such as the French philosopher Nicolas Malebranche (1638–1715), believed in an infinite preformationism. Others, like the Dutch biologist and microscopist Jan Swammerdam, thought that preformationism could be proved by microscopy. In the eighteenth century, other propositions concerning preformation appeared (see Roger [1963] 2015 or Sloan 2002). In the eighteenth century, the hypothesis of preformation continued to be supported by science, religion, and philosophy (Gould 1977)

and was seen as more scientific than epigenesis theory, which required referring to the idea of a "vital force" to explain generation. Moreover, as all the
body organs were already preformed, the embryo only requires the growth
of existing structures and not the formation of new structures ex nihilo. This
developmental process does not require any mystical, vital, or plastic forces
that would "guide" the formation of the organic structures.

However, epigenesis theory would experience a renewal in 1759 when
Caspar Wolff began to analyze the first stages of chick development. He was
surprised not to find in the embryo some of the constituent parts of the adult
animal, such as the beak or the legs, or even organs, which logically should
already appear in the preformed animal. Some embryonic structures had no
equivalent in the adult organism, and he noted the development de novo of
the heart and blood vessels as well as the formation of the intestinal tube
through the folding of a tissue, which was the origin of a distinct structure.
These observations were sufficient to exclude preformationism but were not
sufficient to explain how a new organism was formed at each generation.

To accommodate his findings within a scientific framework, Caspar Wolff
postulated the existence of an unknown *vis essentialis* (essential force), which
implied that the laws that contribute to the organization of embryonic development could be studied and figured out, like gravity or magnetism in physics. His point of view thus differs from that of some of his French epigenesist
predecessors such as Buffon (1750) and Maupertuis (Maupertuis [1745] 2012,
since he adopts a scientific approach in the vein of Newton's classical physics,
which were also in agreement with Christian von Wolff's dynamic conception of matter. However, he does not follow a vitalist approach either—or in
any rejection of Cartesian preformationism—since he avoids using the concepts of "plastic natures" or even the adjective *plastic* from previous philosophical or theoretical views.

At the same time, the weakness of preformationist theory was seen
through its inability to explain crossbreeding and in particular its inability
to explain the fact that couples with black and white skin would give birth to
children whose skin was of an intermediate color. In contrast, Caspar Wolff's
epigenesist theory couldn't completely be embraced because of his reliance
on an unobserved "essential force." Even if the force he described did not have
the same features as the "vital force"—a notion, developed from the eighteenth century by some Montpellier physicians and later theorized by Paul-
Joseph Barthez—Wolff's *vis essentialis* somehow echoed the vitalist views.
However, since Wolff's attempt was clearly seen as an effort to revive the epigenesist theory, the question remains to understand in which way it connects
to the active meaning of plasticity while Wolff didn't employ the term or even

was reluctant to employ it. Previously, I have shown that in Aristotle's philosophy of nature, epigenesis was related to the active sense of plasticity. Does this mean that Wolff understood epigenesis in a radically different way from Aristotle, or does it mean that Wolff refers to the active sense of plasticity without mentioning this term? To answer this question, I compare the use of a plastic- term in the same period by John Tuberville Needham (1713–1781) and the non-use of a plastic- term but of *vis essentialis* by Caspar Wolff.

CASPAR WOLFF'S *VIS ESSENTIALIS* AS THE SIGN OF THE BIRTH OF EMBRYOLOGY

In his work, Caspar Wolff proposed a model of development for both plants and animals based on two key factors: first, the ability of plant and animal fluids (i.e., from living beings) to solidify, and second, the existence of an intrinsic force, *vis essentialis*. Wolf explains thank to these two factors why animals have a heart and plants do not (Wolff 1759). He thought that animals' fluids solidify much more slowly than plants' fluids. The fluids, propelled by the *vis essentialis*, can only form parallel vessels in plants, whereas in animals, they rigidify much less rapidly, eventually leading to the formation of branched vessels (Wolff 1759, 10). Unlike to preformationists of his day, who mostly described what they saw or thought they saw, Wolff attempted to provide a coherent explanation for all the detailed observations of the various developing structures that he made. Wolff was thus one of the first embryologists to focus on the topology of development—the progressive construction in space and time of the embryo. At first sight, this description resembles a mechanical description, as in fluid mechanics. However, the notion of *vis essentialis* nuances this idea because it offers a "principle of sufficient reason" in the manner of Christian von Wolff insofar as it takes into account the existence of necessary constraints that also enable this force to be expressed. Therefore, Wolff's *vis essentialis* is definitely distinct from Georg Ernst Stahl's animism (Roe 1979, 20). While Stahl thought that the origin of life lay in the soul (Stahl 1737), Wolff makes a clear distinction between life processes and the function of the soul. His view implies a separation of "vegetative" processes from both mechanical processes and processes that depend on the soul—such as sensation or thought. These vegetative processes are thus based on both the essential force (*vis essentialis*) and on the identification of constraints on the vegetative processes. The challenge remains to understand what underlies Caspar Wolff's essential force. To clarify his position, he indicates that his *vis essentialis* is different from the *vis plastica* of John Turberville Needham (Wolff 1759; Needham 1747).

Needham's essay clearly shows the influence of Leibniz's philosophy. For Needham, the visible uniformity in nature, beyond the diversity of beings, can only be the result of a preestablished harmony in the universe. By evoking the image of a drop of water being like a sea containing millions of animals, Needham draws on an image from Leibniz's *Monadology* of the pond full of fish (Needham 1747, 2–3). The link between Needham's work and that of the philosopher Leibniz is confirmed by Needham's biography (Murray 2000). Stacey Murray, the author of Needham's biography, points out that the French naturalist Georges-Louis Leclerc de Buffon introduced Needham to some of the ideas of the German philosopher and mathematician Leibniz and that

> Needham not only accepted this idea, but he further believed that when organisms died and decayed, their individual molecules continued to live and could join together to form new living matter. He believed that a force, which he called the "plastic force," brought these molecules together, much like oppositely charged atoms will be drawn together. (Murray 2000)

It seems that if Needham's *vis plastica*, like Leibniz's "plastic natures," refers to a mereological view of the organization of matter and nature, the embryologist Caspar Wolff, on his side, with his notion of *vis essentialis*, examines the organization of matter and nature as a whole. Needham's *vis plastica* and Wolff's *vis essentialis* thus refer to divergent views of the organization of living beings. Wolff's influence on embryology was definitely greater than Needham's. Some commentators have highlighted the innovative aspect of Wolff's work on generation and described it as the first scientific work of embryology in modern times.

> Here appears . . . all the originality of the Wolffian epigenesis. Wolff proposes a model of differentiation that cannot be assimilated to that of Aristotle and Harvey. If he corrects the old epigenesis, it is because "formation takes place through a series of distinct creations of systems (nervous, muscular, digestive, etc.) rather than isolated organs, each of them forming a relatively autonomous whole." Wolff's epigenesis is from the outset those of connective systems and not of organs. (Wolff and Dupont 2003, 50)

The idea of development originating in "connective systems" rather than in isolated organs can be traced to Christian von Wolff's philosophy. Nevertheless, Caspar Wolff's concept of *vis essentialis* is original and goes beyond Christian von Wolff's metaphysical ideas about a dynamic organization of matter since it concerns living systems. Its main characteristics appear when Caspar Wolff responds to the criticism of the most vehement preformationist, Albrecht von Haller (1707–1777). His "essential force" should be understood not as a cause but as an effect of a physico-chemical nature, which, associated with circumstantial and natural causes, contributes to the establishment of the general structure of organisms. The historical opposition between the mechanism and vitalism no longer plays a central role in Caspar Wolff's theoretical approach because he is both sensitive to the Newtonian forces at work in the organization of matter (i.e., repulsion and attraction) and its physico-chemical constituents and to an already "contemporary vitalist" approach that postulates that there is something other than just mechanisms, such as organic emergent properties, to explain the capacity of the organization of living beings. Wolff's original view of a lively debate was only a "way of asserting that organisation exists in the living [being], and that this organisation is not something fundamentally mystical. [impenetrable] and unamenable to scientific attack, but rather the basic problem confronting the [scientist]" (Needham [1968] 2015, 7).[1] Wolff addresses this problem by confronting the explanation of other causes of the organization of living beings: natural and circumstantial causes, which also contribute to the organization of the structure. This research eventually led him to focus on the natural causes of "monsters," which are animals with abnormalities, as well as to the issue of variation (Roe [1981] 2003). Wolff therefore does not stop with the study of the generation itself but pursues his research by applying the study of generation to the understanding of diversity and variation (regular and teratological) in nature. He is therefore the first to have a developmental theory of species diversity based on his concepts of *vis essentialis* and *materia vegetabilis qualificata* (i.e., qualified vegetative matter—the matter that has vegetative qualities).

The *vis essentialis* is reminiscent of the seventeenth-century Neoplatonists' concept of plastic nature because it refers to an organizational capacity understood as an essential principle. But Wolff's methodology—based on observation and description—forces him to reject the "plastic nature" terminology, marked by a certain mysticism because of its use by some thinkers of the previous century who associated vital processes with intellective properties (soul, vital principles).

The two key elements of Wolff's philosophy of science were his rationalism and his opposition to mechanical reductionism. Furthermore, he attempted to avoid the extremes of both mechanism and vitalism, as these were defined in his day, and to create a philosophy of biology of a different caste. (Roe 1979, 12–13). John Needham, on his side used the notion of *vis plastica*, because even though he conducted experiments on microorganisms with new techniques, he remained convinced that these microorganisms come from a spontaneous generation. If the *vis essentialis* appears as a key concept in the history of embryology, the notion of "qualified matter" would open a new chapter in the understanding of plasticity in the life sciences.

WOLFF'S QUALIFIED MATTER AS THE MODE OF
PLASTIC PRODUCTION OF THE SPECIES STRUCTURE

In his treatise *Objecta Meditationum pro Theoria Monstrorum*, Wolff develops
some of his most interesting theories on the relations between the milieu and
the development of form, the nature of the species, and the sources of embryonic organization. He centers his reflection on the distinction he makes
between "structures" on the one hand, and the "cause of structures" on the
other. This allows him to make a distinction between "varieties" and "species,"
which helps to explain how plants, which belong to the same species, can have
different appearances.[2] For Wolff the species is the *cause* of the similarity that
can lead to the difference; the form or the structure are the *effects* of this difference. Moreover, the species remains unchanged under the action of external
conditions, whereas the structure can vary according to external conditions.

In order to clarify his definition of the species, Wolff indicates that the "essential" feature of the vegetative body is secretion. The vegetative substance is
capable of producing particular structures because it has certain "qualities."
These qualities are attributes of the vegetative substance by which the forces
of vegetation are used in various ways. To explain what this notion of quality
means, Wolff distinguishes it from the capacity, the degree, and the mode of
vegetation. Since vegetation (or "vegetative force") is the fundamental process that leads to the production of the general structure of the organism,
vegetative capacity refers to the ability to produce a structure. "Degree of vegetation" refers to the size and quantity of the structures produced, which is
affected by changes in the milieu. The creation of varieties (and of monsters)
depends largely on this degree of vegetation (e.g., extra fingers in humans
attributable to the climatic influence on quantity of vegetative structures). Finally, the "vegetative mode" brings together the qualities of plant and animal
matter. It is through *materia qualificata* (matter, which has vegetative qualities) that the organism produces the specific structures of the species. It is
therefore the vegetative mode alone that must be associated with the species
and not the external shape of the organism.

According to Wolff, no new species can be created from a variation or
a monster because the *materia qualificata* (the vegetative mode) cannot be
modified by changes in the milieu. The standard structure is therefore the
product of the "qualified matter" associated with the usual living conditions.
Conversely, monsters and all varieties are the products of the same vegetative
qualified matter associated with a teratogenic milieu. The development mode
(vegetative mode) of the structure is therefore species specific. This mode
determines the limits that constrain the structure (beyond which it can't be

produced). Wolff conceives the development mode in an essentialist way.[3] Within each mode of development, quantitative differences can be expressed, degrees whose relative expression depends on environmental factors. The expression of these differences gives rise to varieties or monsters (differences in structure but within the limits imposed by the mode, because there is no change of species).

Wolff continues his argument about qualified matter and uses it to explain development. Indeed, in addition to its importance in determining the specificity of the species, the *materia qualificata* is also responsible, in Wolff opinion, for the progressive organization of the embryo. It must be understood that this particular *materia* is not for Wolff an entity that can be located in the embryo; it is rather a disposition (one or more passive qualities) that the living organism possesses. One of the qualities Wolff frequently uses as an example is the ability living organisms possess to solidify their substances. For Wolff the different degrees of solidification determine the most fundamental differences between organisms. Unlike plants, which solidify their substance during their lifetimes, animals, on the other side, only solidify their matter during embryonic development (something that will later be called "primary ossification" in histology[4]). By moving away from both mechanism and vitalism and adopting a conception of the organism as a "qualified substance," Caspar Wolff attributes to matter a role in the generation process that was not possible for his fellow mechanists. This allows him to explain why the embryo develops in the way that he does without referring to any idea of preexisting organization or the invocation of driving forces.

At that time, preformationist theories were based on the assumption common to many seventeenth-century thinkers that matter was simply inert. On the other hand, Caspar Wolff's conception of matter, as well as his view of the "system of the living beings," influenced both Romantic embryology (Goethe's metamorphosis of plants) and above all Christian Heinrich Pander (1794–1865) and Karl Ernst von Baer's (1792–1876) "theory of germinative layers," a fundamental chapter of the history of embryology (discussed below). Even today, one can still see a convergent evolution of Wolff's idea in contemporary embryology. For example, theoretical and experimental work by contemporary developmental biologists such as Ellen Larsen and Stuart Newman, who are interested in the dynamic processes that characterize the mechanisms of development and the organization of living structures, strangely echoes Wolff's reflections (Larsen, Lee, and Glickman 1996; Newman 1992). The only major theoretical difference to note is that what Wolff once called qualified matter is now mostly called plasticity or plastic phenomena.

Indeed, although the term *plasticity* is not used by Wolff in his essays, his *materia qualificata*, the property of matter to be both dynamic and sensitive to environmental variables, will be characterized from the twentieth century onward by the notion of plasticity. Wolff's rejection of the use of the term at the time can be interpreted by his epigenesist leanings. Indeed, while plasticity concepts would have suited his theory well, the use of plastic- terms were, during his time, still very strongly linked with Leibniz and his preformationist conceptions. However, as I will show, plastic- terms reappear in embryology in the nineteenth and twentieth centuries. They would even reappear very often as an explicit reference to the extension of certain ideas introduced by Caspar Wolff himself.

There is also another theoretical current that I have not yet discussed because it is slightly outside the discipline of embryology, but it also inherits from the Wolffian legacy and makes specific use of plastic- terms in parallel with the previous tradition. The philosopher Immanuel Kant (1724–1804) and the biologist Johann Friedrich Blumenbach (1752–1840) are at the origin of this trend, and their works contribute a posteriori to extending our understanding of the theoretical place of the concept of plasticity in the discipline. It is by looking at this ancient theoretical perspective that we can show how the old metaphysical issues, linked with the concept of plasticity, are reformulated.

From Plastic Development to Plastic Organisms: Kant and Blumenbach's Synthetic Approach to Generation

From the end of the eighteenth century, Kant and Blumenbach's ideas independently illustrate the new approaches to generation that combine the "dynamic" epigenesist approach, focused on the "mode of structural organization," and the preformationist approach, based on the "molecules in charge of structural organization." The two authors believed that a synthetic view was possible insofar as there was no incompatibility between the two types of interrogations raised by these approaches. The idea was to reconcile the explanation of what "gives rhythm" to development, the embryonic organization (conceived in a mechanic way), and the explanation of what is at the origin of the organization of the developing embryo. Kant's solution did not, however, eliminate the dichotomy that existed within the field between supporters of neopreformationism and their epigenesists opponents, but it gradually and durably led to a change in the philosophy of ontogenesis.

For Blumenbach, the source of the embryonic organization could be identified in the reproductive material (ovum and sperm), and the organization must be seen as a teleological event (Dupont 2007, 47). Indeed, if the embryo was somehow prespecified, then no vital force, whatever it might be, was necessary to explain the origin of its constituents. On the other hand, it became necessary to refer to a teleological view (the end explains the mean) to understand the reasons for such an organization. Kant, on the other hand, points to a possible reversal of the philosophical question raised by the issue of the origin of the organization (the problem encountered with epigenesis) by showing that development is characterized in some ways by the "achievement" of the organism as a whole that brings the parts together. According to him, the developmental process is carried out by the parts but in relation to the totality as its end. In Kant's work the metaphysical problem of the *origin* or the "starting point of the organization" becomes that of the *purpose* of such an organization. It is no longer about explaining the "origin" of the organization but its "ends" (what the organization leads to: the formation of an organism). The consequence of such a reversal is also the "solution"[5] of the problem of the origin of the organization, a solution that remained incomplete with the epigenesis conception since it was necessary to refer to additional vital forces.

Rather than truly "reconcile" the two opposing theoretical approaches, Kant and Blumenbach's views introduce a new type of philosophical inquiry into embryology that crosses both approaches. This inquiry concerns the best way to apprehend from now on the relationship of the parts to the totality taking into account their temporal distancing.[6] Therefore, the point is no longer just to determine the origin of the organization or the means of that organization but to understand what exactly is being observed when one considers the developing organism. Is it its parts independently of each other, or are these same parts conceived as a whole (internal purpose), that are developing (Huneman 2007, 90)? The study of the progressive organization of the parts is based more and more on what it produces: a mature, adult organism that can in turn reproduce. It should be remembered, for example, that Caspar Wolff did not mention the reasons for fluid solidification when he described the formation of vessels. He was just trying to understand how and why these parts were formed. For Kant, in the parts that organized themselves, it is already possible to conceive of the whole as immediately linked to an organization in the making. It is possible to focus not only on the process itself but also on the process as it produces a complete organism. In other words, for Kant, this idea of totality does not exist in the organisms themselves but as a "principle of knowledge" that allows us to understand development (Kant, *Critique of Judgement*, §65; see Huneman 2007).

The Kantian concept of "original organization" instantiates the idea of a whole as a principle of knowledge in the context of embryology (Huneman

2014). It implies that the actual cause of the parts' arrangement lies in the parts themselves, which are mutually generated—as well as their relations—according to the idea of the whole that will be produced. In other words there is no ontological distinction between the mechanical explanation of processes (epigenesis) and the teleological explanation of the organism's formation (preformation). Between the two, there is only an epistemological distinction concerning the way in which the parts constitute the whole. On the basis of this idea, it becomes possible to combine the two types of explanation within the same theoretical scheme to achieve an empirical synthesis—which was not given in the Leibnizian or Wolffian (Christian von Wolff) approach to ontogeny.[7]

From a theoretical point of view, the main consequence of such a synthetic view is that the focus of study can be not only on the processes (including dynamic aspects of development) or on the entities at the origin of these processes (including the mechanical aspect of development) but also on the objects at the end of the process—the formed organisms, which are all the result of a singular development. From this point on, the study of the plasticity of organisms will become a real topic.

Indeed, one of Kant's objectives through his theoretical approach was to provide a systemic overview of the diversity found in nature (Fisher 2007, 105). Kant's theoretical analysis of generation will contribute to thinking jointly about the two meanings of plasticity in the study of generation: the passive sense (diversity of forms that can be adopted) and the active sense (capacity to develop). Indeed, for Kant, the issue of generation also had important implications for the classification of natural products (Fisher 2007, 110–11). Consequently, while before Kant and Blumenbach the idea of plasticity was used to emphasize the development-specificity of the organism under consideration (its plasticity to be linked to its mode, to its development process), after them the distinction between a mechanical explanation (à la Descartes) and a teleological explanation (à la Leibniz) is no longer relevant. For Kant, development is somehow summarized by its result, and the plasticity of living organisms has to be thought at the final level, those of the whole organism that is the result of a certain composition of its parts in time. This approach allows a comparison between plastic entities, that is, a comparison of the plasticity of animals to plants, the plasticity between different species, or even between different organisms of the same species. What was possible to observe both at the level of the parts and at the level of the processes in the preformationist and epigenesist views of generation can now be seen in the light of its final products.

This approach must, however, be distinguished from the implicit use of the concept of plasticity in embryology (with Caspar Wolff's *matteria qualificata*), which will reappear in embryology in the nineteenth and twentieth centuries but with a more explicit use of plastic- terms.

Plasticity as an Embryonic Disposition

THE PLASTIC EMBRYO AS A SHAPEABLE EMBRYO

Mostly in continuity with Wolff's reflections, in the early nineteenth century the term *plastic* is commonly used among embryologists to characterize the body's "passive ability to adapt" to its "milieu," an ability for organic matter to be "shaped" by this milieu. This view echoes Caspar Wolff's definition of the "degree of vegetation of the qualified matter" and addresses the long-standing problem of the formation of monsters.

But the nineteenth-century authors do not use Wolff's terminology: they prefer the use of the adjective *plastic*, as evidenced (to take just one example) by this quote from Etienne Geoffroy Saint-Hilaire (1772–1844) about Johann Friedrich Meckel (1781–1833): "As for the monstrosities to which Meckel assigns an excess of plastic force, it would be the excessive development of the other parties that would have delayed the evolution of the sexual organs" (Saint-Hilaire 1837, 264, my translation). Contrary to what Saint-Hilaire seems to assert, Meckel does not refer in his texts to the idea of a plastic force in an active sense but rather in a passive sense, in the sense of a passive deformation, resulting from the effects of the milieu. Therefore, Saint-Hilaire's quotation illustrates both (1) the fact that among the authors of his time there is still some confusion about the status of plastic- terms, conceived sometimes in an active way, sometimes in a passive way depending on external constraints, and (2) that Wolff's qualified vegetative matter had a strong influence on embryologists' thinking. Taken together, it appears that plastic-terms were used both to echo Wolff's ideas and to maintain both ideas of a passive and active force at work in developmental processes.

This second aspect is clearly illustrated in this second example of polysemy. In his book on vertebrate development, Martin Heinrich Rathke (1793–1860) uses the adjectives *vegetative* and *plastic* to describe the posterior pole of the oocyte and the embryonic layer associated with it (Rathke 1861, 21). It is this pole that, according to him, is at the origin of many organs. The adjective *plastic* seems to refer either to an active architectural capacity of the tissue that leads to the formation of various organs or to a passive malleability or flexibility of the tissue. It is this passive malleability that, according to him, enables the progressive closure of tubular tissues, especially during the development of the intestines. Rathke's argument also illustrates the gradual transformations in embryological thinking between an animistic view of nature and a progressive empirical analysis of developmental mechanisms where only what can be positively observed and described in anatomical terms is taken into account. Plastic- terms appear with both connotations. In German, *plastisch* appears as a descriptive term, mostly with a passive meaning, used to describe the malleability of the structure.

Furthermore, for many nineteenth-century authors, the adjective is frequently used to describe certain tissues that can be isolated by embryologists. Meckel uses it in his book *Archiv für Anatomie und Physiologie* to describe the nervous system (Meckel 1827, 330–59). He refers in particular to "plastic organs" and "plastic nerves." It is clear that the adjective *plastic* is used there to designate "something soft," assimilating the embryonic property of plasticity to a strictly physical property (like the irreversible deformation highlighted by the philosopher of nature and scientist Robert Hooke [1635–1703] a century earlier by the law of elasticity[8]). Plasticity is understood here as a passive physical property not specific to living bodies, plastic deformation being linked to the physical stresses on tissue or organs. This somehow adds a third or even a fourth meaning (if one counts the vitalist or animist meaning still driven by the term) to be linked with plastic- terms in the embryology of the nineteenth century.

This diversity of meanings for plastic- terms can also be explained by the fact that at the beginning of the nineteenth century, embryology was not yet a field of study distinct from anatomy (if it ever became so). Embryologists mostly identified and compared structures according to their shape and material. Therefore, their challenge was not yet to explain the origin of the different tissues. Meckel, whose theories are strongly influenced by *Naturphilosophie*[9] (one of the famous representatives of this trend is Friedrich W. Schelling), states that higher animals pass, during their development, through stages corresponding to the adult forms of lower animals (on a phylogenetic scale) (Meckel 1830). Such an idea was taken up and popularized later by Ernst Haeckel (1834–1919) with the famous formula, "ontogenesis recapitulates phylogenesis."[10] The formula would remain famous even after the first publication in 1859 of Darwin's *On the Origin of Species* (see Gould 1977). This particular example illustrates that embryologists' interest at that time was quite different from seeking to explain the specific origin of tissues or their progressive organization, something that would become the main concern of their successor developmental biologists.

In such a context, for embryologists of the nineteenth century influenced by the transformist movement, biological plasticity appears mostly as a passive property of living organisms directly influenced by the effects of the milieu on living structures. Embryology of this period would emphasize the idea of plasticity, as "the ability to adopt a number of varied shapes." Nevertheless, they do not neglect the active sense of the term, even if this aspect is not at the center of their concerns at the moment.

However, Caspar Wolff's work, which remains strongly influential, requires a dynamic feature to explain the progressive organization of the embryo in formation, a feature that relies on the properties and qualities of the material (a vegetation mode) itself specific to the species and structure adopted. It is this dynamic feature that has to be linked with the active meaning of plastic- terms. Therefore, after Wolff, and for the nineteenth-century embryologists, any reference to plastic- terms in embryology implicitly means both the passive sense of plastic- terms with the idea of malleability of the developing tissue and the active sense of plastic- terms with the idea of an architectural force of organic matter. Wolff had first undertaken the empirical investigation of the causes of the processes of organizing matter, the results of which would require nineteenth-century embryologists to link the adjective *plastic* in its active sense with the scientific context of the time. It was something that was not possible in the seventeenth and eighteenth centuries, when plastic- terms were used by philosophers of nature. Therefore, even if a "vitalist" connotation remains linked with plastic- terms until the nineteenth century, and even if they had, for a time, disappeared from the terminology of embryologists (and first and foremost from Caspar Wolff's work), they are gradually reappearing in embryology in an increasingly experimental context, linked both with a passive and, increasingly, active meaning.

THE "PLASTIC NATURE" OF THE EMBRYONIC TISSUES

From the second half of the nineteenth century, concerns about the material causes of development—which had been ignored by descriptive embryology in favor of a purely anatomical description of the different stages of development—started resurfacing as a result of advances in physiology (Dupont and Schmitt 2004).[11] Experimental embryology, later called "mechanics of development" [*Entwicklungsmechanik*] by its founders, was born. Defended by Wilhelm Roux (1850–1924) and Wilhelm His (1831–1904), this new study of development became dominant in Germany and in the United States from the late nineteenth century onward. Through meticulous experiments, biologists test the influence of variations in physico-chemical conditions such as gravity, pressure, temperature, or chemical concentrations on the development of carefully selected model organisms (mainly amphibians and marine invertebrates) that are easy to handle. His and Roux are mainly interested in the implementation of theoretical models: brain models for the former, and models of the different regions of the ovum for the latter (His 1887, Roux 1888).

For most embryologists who follow His and Roux, plastic- terms are never used to mean "something soft" in a tissue but always have an active meaning, in the sense of a potentiality in becoming. Several authors would even attempt to clarify the causes of this potentiality—today often called morphological plasticity[12]—somehow pursuing the efforts Caspar Wolff had begun a century earlier with the characterization of qualified matter.

Among Cambridge and then Oxford embryologists, we observe an effort similar to that of their colleagues on the continent. Adam Sedgwick (1854–1913), a student of Francis Maitland Balfour (1851–1882), tries to specify what causes the possession of a "plastic nature" during development—taking up the term used a century earlier by the Cambridge Platonists (More and Cudworth). In attempting to explain the development of the three embryonic layers,[13] Sedgwick notes about the ectoderm, "The layer was possibly of a plastic nature, and allowed the protrusion of the central mass at one or more points" (Sedgwick 1889, 115). The "plastic nature" that characterizes the embryonic layer will allow Sedgwick to identify, for the first time, a phenomenon of development, which will be then called epiboly.[14] The particular cause of this developmental mechanism seems to depend strongly, according to Sedgwick, on this "plastic nature" (or plastic property), which characterizes the first few layers of the embryo. This use of the term with this meaning shows that the use of plastic- terms is no longer banned from the terminology of embryologists: it seems also definitively freed from any "vitalist" or animist connotation.

However, despite the fact that all these authors, pioneers of experimental embryology, seek for the first time to clarify what it means to be plastic for a tissue or for the embryo, the use of plastic- terms still seems to point to a problem more than it brings a real solution to its understanding. A new step will be taken with the development of the cell theory.

CELL THEORY: MOSAIC EGGS VERSUS REGULATIVE EGGS

In the early 1890s embryologists based their thinking on specific model organisms, which were the eggs and embryos of marine invertebrates (ascidians, sea urchins, starfish, etc.) (These new model organisms are still essential and central in embryological research today.) Their investigations develop thanks to the development of maritime stations that welcome embryologists and provide them with adequate resources (see Fischer 2002a).[15] In addition to this marine resource, embryologists are also increasingly using amphibian eggs, which would replace, for a time, the hen's egg because they appear to be a more suitable material.

In 1887 Laurent Chabry defined and demonstrated the anisotropy of the ascidian[16] egg: this egg cannot repair the damage induced by the experimenter if one of the blastomeres (the first embryonic cells) is destroyed. Influenced by physiologist Eduard Friedrich Pflüger (1829–1910) and his experiments on early amphibian development, Roux carried out similar experiments, but his experiments allowed him to obtain half embryos (i.e., embryos cut in half) of frogs at different developmental stages. Roux thought that the amphibian egg was "like a mosaic of elements, each of which would correspond to parts of the future embryo" (Roux 1888). The name of his theoretical model—the "mosaic egg"—was taken from this observation. Gradually, this type of experiment and observation was extended to other organisms, particularly to sea urchin eggs. Hans Driesch (1867–1941) concluded, from an experiment he carried out in 1891, that there is an isotropy of the sea urchin egg. During the experiment, he noted that the removal of a blastomere at stage two, four, or eight blastomeres did not interfere with the further development of the sea urchin. Oscar Hertwig (1849–1922) continued these experiments and used his observations to formulate a theory of development through epigenesis. Actually, this "epigenesist theory of development" came to be called *neoepigenesis* by later commentators because it no longer resembled Aristotle's epigenesist theory or that of Wolff, although it was part of the same tradition (Fischer 2002b).

Driesch described the egg being divided as "a harmonious equipotential system." Each cell contains, in latency, the potential to produce a complete organism. Differentiation occurs because the forces surrounding blastomeres vary according to the differences in the original spatial and temporal positions of these cells. Driesch thus succeeds in obtaining complete larvae from blastomeres from the four-cell stage, separated from the rest of the sea urchin embryo. From these observations, he showed that the early embryo of certain species (formed by a few blastomeres) had a power of "regulation." Consequently, as the notion of "mosaic egg" proposed by Roux seemed obsolete to him, Driesch proposed the concept of "regulation egg."

While his epigenesist theory will be seen as different from Caspar Wolff's theory, his work reflects the influence of Caspar Wolff but with the difference that the scale of observation differs. The organism is no longer seen as an indivisible whole in which structures are created that carry out certain functions—as was the case for Wolff—but one considers, from the nineteenth century and the birth of cell theory, that "every living being is constituted by units that have a certain autonomy and are capable of dividing themselves, the cells" (Dupont and Schmitt 2004). The level of observation becomes cellular. But it does not necessarily lead the authors concerned to adopt a "reductionist" point of view in the sense that *only* cellular mechanisms would account for the phenomena of generation.

For Driesch, in particular, the opposite situation is observed, and he will finally embrace a vitalist point of view. Vitalism had persisted in Germany because of the progress of preformationist theory (also called evolutionary theory[17]) and because of the Haeckelian method, which had prolonged this inclination, favoring phylogenetic explanations over explanations by proximate causes. At the beginning of the twentieth century, Driesch became an ardent supporter of vitalism and the Aristotelian notion of entelechy (life in the making). However, Driesch, who was quite faithful to the terminology and foundations of the field established by Wolff, never referred in his work to any plastic- term.

In contrast, neopreformationists[18] believed that the destiny of embryos was fixed, once and for all, in the fertilized egg. According to the mosaic theory of development, the egg was as well specified and structured as the adult (Roux 1888). It was divided into regions that were intended to produce the specified parts and organs of the animal as a whole. Ontogenesis was thought of as the deployment of a predetermined structure (Jenkinson 1909).

Cell division was the previously undiscovered process that helped to explain the gradual appearance of the body's shape. Therefore, from the moment this phenomenon was discovered, it was no longer necessary to refer to any occult powers or plastic forces. In a way, both the Cambridge Platonists' view and Leibniz' view started to go out of fashion. Continuing these observations, Wilhelm Roux and August Weismann (1834–1914) argued that the nuclear divisions during cleavage (soon to be called meiosis and distinguished from mitosis) were "unequal" because the determinants of future structures were divided during blastomere cleavage. Roux provides a first empirical confirmation to his doctrine when he succeeds in obtaining a half embryo after destroying one of the blastomeres of the frog egg cell at the two-cell stage. He also observes that in some situations, when the two blastomeres are simply separated, he manages to obtain two whole embryos.

If in the 1890s there was a tangible theoretical continuity between the new theories (neopreformationism and neoepigenesis) and the older ones, it would be wrong to underestimate the differences, which are considerable, both for theoretical and contextual reasons. For instance, the two new conceptions—epigenesis and preformation—are no longer so radically opposed. Differences persist, but mainly between a focus on the mode of structural organization of the embryo (neoepigenesis) and a focus on the proximate causes of development (neopreformationism). On the side of the neopreformationists, the theoretical continuity with the old conceptions is much less clear than on the side of the epigenetists. Nevertheless, one of the central and recurrent elements is that preformation theory is not confronted with the difficult question of how the embryo manages to reach such a level of complexity in its organization during its development. The mode of structural organization of the embryo is already "incorporated," "preprogrammed" into the egg. It is therefore possible to explain development mechanically without the risk of falling into a form of materialism. From the scientific point of view, this approach was finally considered more satisfactory at that time than the epigenetic approach—even if the scientific content of the original preformationist theory proved to be wrong (the embryo does not already exist in a very small size in the egg)—because it can provide an explanation for development. For this reason, the approach underlying this neopreformationism—which consists in reflecting on the gradual implementation during the development of an already preestablished material—acquired, as early as the 1890s, a significant importance among scientists.

In line with this school of thought, nineteenth-century embryology focused essentially on identifying the proximate causes, on the search for mechanical explanations (such as *Entwicklungsmechanik*), despite the initial success of the theory of epigenesis when Wolff had revived it in the eighteenth century. After Wolff, the theory of epigenesis was replaced by concepts of a vitalist and teleological nature, such as those of Blumenbach's "vital force" or *Bildungstrieb* (formative tendency). The questions about the origin of form and variation that Wolff had studied, and that referred to the "active" sense of the term *plastic*, are no longer the focus of research in the study of development in the nineteenth century.

This change in the theoretical interests of embryologists seems, at first sight, to be linked to the development and elaboration of new scientific theories (cell theory is the most striking example) that in turn are correlated with the improvement of investigation methods for the developing embryo. Different and increasingly precise scales of analysis are emerging. Jane Oppenheimer pointed out that if, for a very long time, the embryo was studied as a whole (Aristotle, Malpighi, Wolff, and others), it would then be studied according to its layers (Von Baer, Pander, His, and others), then according to the relationships between these layers and the cells that constitute them (Chabry, Roux, Driesch, Spemann, and others), and finally according to its cellular compounds (Hertwig, Boveri, Wilson, Conklin, and others) (Oppenheimer 1967). This "experimental revolution," by allowing nineteenth-century biologists to gradually conceive the existence of different underlying organizational levels, favored the growth of a vast field of research on the proximate causes, always better identified and specified without totally eliminating reflections of a more metaphysical nature. Indeed, the Haeckelian conception (the recapitulation law) exerts a decisive influence on embryology until after the publication of Darwin's *On the Origin of Species*, which switches the focuses of embryological research to the evidence of phylogenetic relationships between species.

However, among embryologists who distanced themselves from this "line" and whose influence was therefore less important than that of their contemporaries, the Wolffian legacy—brought up to date by the discoveries and methods of the nineteenth century—continued. One such example is the work of embryologists Bergmann and Leuckart, who developed a program for a "new embryology." Using the latest research on cell transformation, they looked for ways to determine the efficient causes of organ disposition during development. They called their approach, which is almost like an experimental manifesto, "a physiology of plastic" (eine Physiologie der Plastik) (Bergmann and Leuckart 1852, 36). Few, however, embarked on that original path. Oppenheimer cites the case of the French physician, zoologist, and talented embryologist Dominique Auguste Lereboullet but also the German embryologist Wilhelm His, who in his mechanical explanation of development offered a "physiology of the plastic" (Oppenheimer 1936). He was, at the time, one of Haeckel's fiercest opponents. For this reason, the work of these two authors was not "in vogue," and they encountered little success in their approach mainly because of the absence of a favorable theoretical background.

Between 1850 and 1900, there was a resurgence of the concept of plasticity. Plasticity in the life sciences was no longer conceived solely as a symbol of an architectural force or a property of life. When embryologists use the adjective *plastic*, they no longer rationally explain "why" development occurs but rather try to explain empirically and experimentally "how" it occurs. While metaphysical considerations in embryology seem to concern mainly the relationship between ontogenesis and phylogenesis, some of them also concern the mechanisms and the proximate causes of development.

At the same time, "matter" loses its metaphysical dimension. Embryologists then referred, without any difficulty, to the notion of "plastic material," which appeared as a physical qualification among others (solidification, aggregation) to characterize different observed phenomena of ontogenesis. Plasticity became equivalent to malleability, deformation capacity, sensitivity to environmental disturbances, and there was no need to specify the origin of this potentiality.

While in the nineteenth century, plastic- terms were not yet central to theoretical analyses of development, the study of the embryo, according to the different scales of structural organization of the biological world, raised new theoretical and philosophical issues, such as what structure would confer a plastic character or condition. Could it be the tissue, the cell, or an emerging property linked to the interaction of these different levels of organization? We observe a shift from issues related to the "causes of development" to issues related to the "causes of plasticity."

Investigating the Causes of Plasticity in Embryology

INDUCTIVE PHENOMENA AND SCALES OF ANALYSIS

The preformationist theoretical foundations—the search for the proximate causes—as well as the emergence of new methods of analysis, at increasingly smaller scales, gradually lead embryologists to look for the origin of plasticity and/or malleability that one observes in the developing embryo to a more and more precise cell level. It is from Hans Spemann's (1869–1941) work in the 1920s that the issue of an embryo's plasticity became a highly theoretical issue of embryology (Spemann 1938). In the course of his research, Spemann performs so-called heteroplastic[19] transplants of embryonic territories from one region to another (he transplants the region of the dorsal lip of the frog blastopore[20] into its ventral region). Through his experiments, he tries to identify regions, that is, groups of highly targeted cells, that can develop independently of a specific environmental context. The term *plastic*, used to characterize these cells, combines both an "active" meaning of the term as a capacity to develop and a "passive" meaning as a diversity of forms that can be adopted.

In these experiments Spemann, whose work was largely influenced by cytology and physiology[21]—fields that were undergoing significant development—interferes with the normal physiological development of the embryos he is studying. He used a human hair to cut the embryos into distinct regions. Depending on the position of the blastopore[22] in relation to the cut, he managed or not to obtain embryos that completed their development normally. On this basis, he demonstrated that the blastopore is essential to differentiation and that the moment when the cut is carried out is just as essential—the beginning of gastrulation corresponding to the so-called crucial period of axial differentiation.

In 1923 Spemann and his student Hilde Mangold revolutionized embryology when they discovered a particular region of the embryo involved in the
emergence of the central nervous system. Together they showed that when
this region, the dorsal lip of the blastopore, is transplanted from a salamander
embryo into the future ventral region of another salamander embryo, the latter
invaginates entirely (which means that it continues its normal development).
In addition, the transplanted dorsal lip tissue (and only this region) leads to
the formation of a new notochord that generates a new neural tube and, ultimately, a secondary embryo (Spemann and Mangold (1923) 2003). They called
this region the "organizer" of the embryo. The organizer itself does not contribute to the neural tissue. It forms the pharyngeal endodermis and the dorsal
mesodermis (notochord and somites), which are found under the cells that are
responsible for the formation of the central nervous system. Neural tube cells
come mainly from the host's ectoderm. Therefore, the "organizer" itself was
not seen as a "plastic" region (since it does not have the ability to "produce" a
new type of tissue), but it could "induce" the formation of a secondary dorsal
axis from the tissues of the host. This first phenomenon of induction then trig-
gers a cascade of other inductive events leading to the progressive "construction" of the embryo, hence the use of the term *primary embryonic induction*
to describe this event (see chap. 6 about the transformations of this notion in
contemporary biological literature). This means that a decisive "informative
message" is transmitted through this cell zone and that the message leads to
the generation of an embryonic structural organization de novo.

But what is the "substance" at the origin of this message that produces a
new embryo? Until the late 1930s, most of the work focused on finding the
chemical nature of the inductive signal. But this research was unsuccessful
because many substances were found to be inductive (Dupont and Schmitt
2004, 217). This illustrates the main problem embryologists have to deal with.
Any substance seems to be able to play the role of an inducer. In order to determine the specificity of the developmental process, it would be necessary to
identify some limits, constraints that would explain why development is not
a simple generalized "transformationism."

Pieter Nieuwkoop's (1917–1996) work is a good example of an attempt to
establish "boundaries." He carried out experiments in which he gradually ex-
tracted the cells of an embryo at the end of the segmentation phase in order
to observe the smallest quantity of cells necessary for gastrulation or neurulation to be achieved. His approach appeared to be an attempt to determine
what the smallest existing plastic unit is, the unit below which there is no
longer plasticity, that is, no longer any potentiality to induce a primitive developmental process.

Nieuwkoop's contribution, with the later identification of a second induction stage at the blastula stage, happening before the Spemann organizer and helped pave the way for molecular induction analysis (Nieuwkoop, Johnen, and Albers 1985). His pioneering work also enabled the raise of developmental biology as an autonomous field within biology. As one moves toward the molecular level, genetics become more important in developmental studies. Embryologists no longer saw genetics as a "peripheral" field of study but as a central discipline for progress in embryology.

Conrad Hal Waddington (1905–1975) was one of those embryologists who was convinced by the truth and importance of Mendelian genetics. Waddington first worked in Spemann's laboratory and later in Thomas Morgan's (1866–1945) laboratory in California. During the 1930s he collaborated in Cambridge with embryologists Joseph Needham (1900–1995) and Jean Brachet (1909–1988), and he tried to identify the chemical causes of induction in the Spemann organizer. Together, these scientists performed experiments using many chemicals but without conclusive results.

As molecular signals were observed, biologists were increasingly aware that the complexity of the induction phenomenon requires much more than just identifying these signals. The successes of Mendelian genetics and those subsequently achieved by what will be called "developmental genetics" cannot, even retrospectively, mean that embryological and biochemical research on induction was merely a research step awaiting a transition to molecular analysis. Rather, a fundamental shift happened at this moment, a shift that persisted in embryology: the concept of "developmental potentiality" (linked to the active meaning of the term *plastic*) appears as something that cannot be reduced to a simple structure or a single material entity.

THE SEVERAL USES OF PLASTIC- TERMS
IN THE HISTORY OF EMBRYOLOGY

The discovery of the primary inducer by Spemann in 1924, a major moment in the history of embryology, was a challenge for all subsequent embryologists who worked to identify the origin of inductive information through a close-up examination. Indeed, from the nineteenth century onward and with the spread of cell theory, one of the great ambitions of embryology was to understand morphogenesis at the cellular level. One moves from the observation of the organism as a whole to the identification of tissues, groups of cells, and finally the molecular determinants that enable generation of the organism's development. All these different scales of analysis had consequences for plasticity: plastic- terms were used with different meanings. For instance, in the nineteenth century, the first cellular research on the origins of cancers began. In this perspective, the term *plastic*, appears to distinguish some cancer cells from "normal" cells, as in the case of "fibro-plastic cells,"[23] where the cells seem degenerative with the ability to divide easily because of the unusual density of fibers within the cell (definition in Laurence 1855).

This willingness to understand morphogenesis at the cellular level reappeared throughout the twentieth century, in particular with Viktor Hamburger and Jane Oppenheimer in their studies of cell behavior during development. Oppenheimer noted, for instance, that by changing the environment of certain cells during their development, their cellular fate can be altered:

> When cells are isolated in culture in some synthetic medium, the direction of their differentiation presumably may be influenced by the reaction of cells to factors in the medium: if they are transplanted to a new cellular environment, it will be conditioned in relationship to factors emanating from neighboring cells, and therefore by mutual interactions between cells of graft and cells of host. (Oppenheimer 1955b, 35)

Similarly, it has been observed that cells that usually differentiate into muscle cells can differentiate into cartilage or bone cells if they are subjected to certain environmental factors. To observe such a phenomenon, the cells selected must have a certain "degree of plasticity"; that is, they must not be yet "determined" or induced. This is the case for cells that will be studied and identified, from the 1950s onward, under the name of "stem cells."[24]

During the nineteenth and twentieth centuries, the embryologists' view of plasticity in the biological sphere changed considerably. The adjective *plastic* in French and English was used initially in a relatively vague and indeterminate way, and sometimes with a seventeenth-century animist or vitalist view. By the end of the eighteenth century, plastic- terms disappeared from the writings of experimental embryologists. However, plastic- terms would gradually reappear during the nineteenth century. In the work of embryologists, plasticity was later associated with a more physical passive meaning, referring to a certain malleability or elasticity of living tissue. The rise of cell theory and the shift from embryologists' metaphysical concerns to phylogenic issues contributed to the gradual loss of the old vitalist connotations of plastic- terms. The concept of plasticity can then acquire a new, more scientific meaning, such as the "potentiality" of cells to differentiate. The carrier of plasticity in embryology becomes the cell, whereas previously it was either living matter or the organism. The concept of plasticity acquired a strictly epistemic status in a context where embryology tended to become more specialized. The "plasticity" of cells referred to their indetermination, their potentiality, contrasting with the differentiated, determined cells of the adult organism. By extension, "plasticity" now referred to the embryonic organism in development distinct from the adult organism. At the same time, the term also tended to be used to highlight degrees of "plasticity" between species, that is, their respective ability to be affected by and to react to environmental variations.

Conclusion: The Consolidation of the Concept of Plasticity within Embryology

The history of embryology helps to establish a bridge between the different philosophical points of view that, until the seventeenth century, have made inquiries about the objects of nature. But it also helps bring them closer to the later theoretical thoughts of the first embryologists who, since the beginning of the eighteenth century, have observed and described the developmental process in an increasingly precise way. The opposition between two views on generation and development—epigenesis and preformation—has remained through the history of embryology. However, the theoretical, methodological and technological progresses within embryology have contributed to a progressive and quite considerable change in the embryologists' understanding and definition of these two major theoretical options.

While some fundamental differences have emerged between past and present conceptions, ancient theories continue to strongly influence later embryologists. This close link between a philosophy of generation and its analysis—which has become more scientific—points to the absence of a clear break between, on the one hand, the theoretical thinking of generation and, on the other hand, a reflection based on the empirical observation of generation as a phenomenon. The analysis of Caspar Wolff's work (considered as the "father of embryology") is, in this respect, paradigmatic. Thus, when Wolff refers to the term *essential force*, using the Latin expression *vis essentialis*, the relationship with the *vis plastica* of his predecessor, William Harvey, is clear. Harvey had used the notion more than a century before Wolff to explain generation through epigenesis. Yet if a filiation between the two authors is obvious, Wolff's notion of *vis essentialis* is nevertheless not a plastic- term. How do we explain why Wolff did not use a term that was so much in fashion at that time? Why did he need to create a neologism? His reluctance to use the specific term *plastic force*—a notion that had acquired, since the formulation by the Cambridge Platonists of the "plastic nature(s)," a new connotation linked to the rejection of the Cartesian mechanism—is based on his methodological and ontological commitments. Wolff based his theory of generation on a new methodological context reinforced by a solid empirical observation distinct from the animist or vitalist thinking of the time. In the eighteenth century, the consequence was the temporary disappearance of plastic- terms from embryologists' terminology.

Nevertheless, the different meanings and general definitions of the terms remain with other words. In particular, the active sense of the term *plasticity* (ability to develop a form) continues to be discussed by embryologists who examine the proximate causes of the form's development (morphogenesis). Even if the term disappears, implicit theories of plasticity remain. In addition, the temporary disappearance of plastic- terms among eighteenth- and early nineteenth-century embryologists comes with a major change in the meaning of the term when it reappears because its seventeenth-century philosophical connotations tended to fade away. Therefore, when it reappeared sporadically at the beginning of the nineteenth century in the writings of experimental embryologists, it is henceforth linked to a simple physical quality of the developing material.

However quickly, the reappearance of plastic- terms introduce new interrogations into the structure of living organisms and what is at the origin of the plastic "character" or "condition" of the developing organism. Is it the whole organism, some primordial tissues, or only some delimited cells that are plastic? In the nascent nineteenth- and twentieth-century embryology, there was a need to refer to plastic- terms in an active sense, as a "developmental potentiality," to reflect specificity of the developmental process. This is illustrated in 1923 with Spemann and Mangold's major discovery of the "primary inducer." The very notion of "inducer" takes up—through a notion, *scientifically delimited*[25]— all the dispositional implicit features that are specific to the active sense of the term *plastic* (we return to this notion of "inducer" in chap. 6). The inducer refers to a group of cells that emits signals and interacts with another group of cells to influence its differentiation and "determine" it (specify it in a conditioned manner). It is impossible to identify otherwise than in retrospect the "inductive" nature of the group of cells. It is around 1850 that the term *plastic* was definitively introduced into the vocabulary of embryologists, and we find it today frequently associated with studies that focus on "developmental potentiality" at the cellular level (particularly in the study of stem cells).

But in parallel with this first current of thought about plasticity within the field of embryology, one must also describe a second current, of equally importance, based on the fundamental theoretical upheaval initiated by Kant and Blumenbach at the end of the eighteenth and the beginning of the nineteenth centuries. By seeking to combine the embryonic organization and the material factors on which this organization is based, Kant and Blumenbach rephrased the debate about generation. They put their focus not on the developmental processes (as was the case for epigenesis) or on the developing entities (as was the case for preformation) but on the objects that are the outcome of the development process—organisms.

For Kant and Blumenbach, thinking about the specificity of embryology implies connecting it with the fundamental issue of the diversity of human races, first of all, and, for Kant, with the problem of the difference between racial diversity and species diversity (Huneman 2007). Species diversity is also an expression of the "plasticity of living organisms" (in a passive sense here since it is revealed "afterward," "eventually" by the diversity of the process that had led to it). From that moment on, the second sense of plasticity, the passive sense (as "the ability to adopt a number of varied forms"), until then more neglected, appeared at the forefront of the scene to take root in the life sciences. If, thanks to the Kantian teleological approach to generation, a theoretical problem specific to embryology (such as that of the split between an epigenetic and preformationist conception) seems to be solved, the synthesis of the two senses of plasticity seems far from being achieved.

With Kant and Blumenbach, the "active" sense of the term seems to be eclipsed because the issue concerning the diversity of the living organisms is put forward, in the Kantian view, over the issue of generation itself. However, as I have shown, the active meaning of plastic- terms has an important role in the theoretical terminology of embryological analysis. Therefore, the Kantian approach seems to have its limits, since it does not allow, in its empirical application, to account for all the phenomena of generation. If Kant's view appears to be quite an innovation in the way it helps solving the problem of generation, it brings back old phantoms of preformationism concerning the understanding of plasticity. Indeed, one consequence of Kant and Blumenbach's position in their attempt to achieve a synthesis is the neglect of the temporal and historical dimension found in the active sense of the concept of plasticity. Consequently plasticity—understood as diversity—became mainly a property of adult organisms.

From an ontological point of view, there is a separation between the idea of plasticity as referring to an "incompletion," to an "object that has not undergone any differentiation"—such as stem cells—and the idea of plasticity as a "diversity" between organisms, which will, after Darwin, be interpreted as the result of adaptations. While today a causal physical approach to plasticity in the study of development persists through mechanistic explanations of development (research of plastic tissues, plastic cells, and ultimately plasticity genes), a more systemic approach—similar to Caspar Friedrich Wolff's "qualified matter" (dynamic approach)—seems mostly abandoned in favor of the

Kantian teleological approach to generation that leads to a unique account of the relative plasticity of organisms.

From the end of the nineteenth century onward, a common ground between experimental embryology and nascent genetics was established (see chap. 3). Researchers were interested in understanding the relationship between units of heredity and the mechanics of development. This involved understanding the biological determinism by which fertilized eggs give rise to complex organisms. The debate then opposed the supporters of the decisive role of the cytoplasm (Pflüger, Bischoff, His) and those of the chromosomal heredity (Hertwig, Roux, Boveri). With his research on *Drosophila*, Thomas Morgan slowly switched to the chromosomal hypothesis and soon formulated the gene theory, which then closely linked the history of embryology to that of genetics (Morgan 1917). Apart from these poor links with embryology, genetics gradually grew as an independent field of study to bypass embryology. If the fundamental theoretical problem of embryology was the relationship between epigenesis and preformation, the theoretical problem specific to genetics will essentially be that of the causal evaluation of genetic factors in the determination of living organisms. In the next chapter, I look at how the issue of plasticity became part of the issue of the determination of traits based on genetic data.

The Emergence of an Operational Concept
of Plasticity in Genetics

Once Mendel's theory had become accepted, other biological problems concerned with reproduction, metabolism, development, and the acquisition and maintenance of a specific form had to be reassessed and restated, often with obvious reluctance on the part of those who had reached their views before Mendelism troubled the waters. Animal embryologists were especially disturbed by Mendelism, for their chief problem was to find ways of explaining the integration of a complex set of cells, tissues, and organs into one functioning whole. Mendelism, at first glance, offered only a means of resolving the organism into separable unit characters.

LESLIE DUNN ([1965] 1991, IX)

If genetics is the discipline that has seen the greatest and the most rapid advances in biology since the first half of the twentieth century, it is also the discipline that has had the strongest impact on biology. The impact took place at multiple levels and in many ways. Although it remains difficult to say exactly who (August Weismann, Gregor Mendel, or Wilhelm Johannsen) made untenable the belief in the inheritance of acquired characters—a belief that had solaced most nineteenth-century biologists—a shift remained patent (Dunn [1965] 1991, ix).

In the nineteenth century the intent had not been to understand the orderly development of structures on a geological time scale; instead, it had been to understand the perfect adaptation of the observed forms to their functions. Such perfection was often interpreted as the consequence of some form of inheritance of acquired traits for which organisms were considered to be "positively reacting to their environment" (Loison 2010). However, such an interpretation gradually disappeared with the rise of genetics.

The use of the concept of mutation, for instance, illustrates the kind of shift I intend to highlight here in relation to the idea of plasticity. From its first use in paleontology in the 1890s, when the term *mutation* was understood to mean the slightest change from one form to another (e.g., a change in morphology of fossils belonging to the same lineage; Waagen 1886), to its later use in genetics, where *mutation* came to mean a sudden change that gave rise to new "types" or species, major conceptual shifts could already be identified (Dunn [1965] 1991, ix). Similarly, the emergence and expansion of genetics also had a major impact on how biologists understood the concept of plasticity. Similarly with the word *mutation*, although it is still unclear when the transition took place from a *certain idea* of plasticity, manifested in various ways throughout the history of embryology (see previous chapter) to the formulation of a precise definition of the notion of plasticity in the context of the emerging genetics, the change is nevertheless patent. The new science of genetics would give birth, in the mid-1960s, to a concept of plasticity defined in an operational way through the formulation of the notion of "phenotypic plasticity," which can be defined, to a first approximation, as the ability of an organism to express different phenotypes for the same genotype according to the environmental conditions. The definition itself refers to other important concepts of genetics (i.e., "phenotype" and "genotype").

In order to trace the origins of this change, in this chapter I first examine the contexts in genetics in which the first considerations of the relationship between organisms and their environment emerged. In doing so, and for the sake of clarity, I distinguish between the uses and meanings of the term *plasticity* in genetics on the one hand and its uses and meanings in evolutionary biology on the other. However, since it is impossible to talk about plasticity in genetics without making some preliminary remarks about evolutionary biology (the uses of "plasticity" in evolutionary biology will be developed in the next chapter), the first section is devoted to some introductory remarks about the relationship between genetics and evolutionary biology and the specificities of the former with respect to the latter, and of both with respect to plasticity. The second and third sections describe and analyze the emergence and gradual transformation of the concept of "norm of reaction," which was first used to illustrate environment-genotype interactions before the notion of "adaptive norm of reaction" and its link to evolutionary processes received full attention. I argue that the concept of phenotypic plasticity as it is used in contemporary biology has inherited these earlier notions and that most of our misunderstandings of plasticity processes are based on earlier difficulties raised by the concept of the norm of reaction. The fourth section then examines in more detail Richard Goldschmidt's distinction between "norm of reaction" and "norm of reactivity" and his attempt to link genetics to development. Indeed, based on what may appear to be historical episodes in biology, I show that notions of plasticity such as that of the "norm of reaction" became lynchpins of reflection in biology in general, especially when they link issues of developmental biology (which will become those of genetics) with those of evolutionary biology.

Distinguishing Genetic Factors from Environmental Factors

While Mendel's laws had provided an excellent explanation of discontinuous variation between generations at the time of their rediscovery in 1901 and for several decades thereafter (e.g., between 1900 and 1930), opposition to their systematic application to all living phenomena had persisted (Mayr 1982, 777). Opponents did not deny the existence of Mendelian inheritance; they denied that *all* inheritance should be considered Mendelian in nature.

As early as 1901, species inheritance (i.e., the inheritance of the "qualities" of the species) was of particular importance to Darwinian zoologists and botanists in the initial context of a theory of evolution by natural selection. However, when these same zoologists and botanists began to take an interest in the new discipline of Mendelian genetics, they did so mainly through the writings of Mendelians who also spoke of evolutionary biology, such as Hugo de Vries (1848–1935) or William Bateson (1861–1926). These writings seemed somehow unacceptable to them because they described the discontinuous nature of inheritance as evidence for the discontinuous nature of the appearance of species. Darwin had advocated a gradualist view of evolution in which small continuous variations predominated. The result was the emergence of two opposing camps that were to remain in competition for a long time: on the one side, the Mendelians who tended to think in essentialist terms and focused on the behavior of elementary units of heredity; on the other, the Darwinians, who were concerned with population phenomena and focused on interpretations in holistic terms.[1]

When Darwin formulated the theory of natural selection in *On the Origin of Species* in 1859, he also gave specific definitions to many colloquial terms such as *adaptation*, *variation*, and *evolution*. From then on, these terms were analyzed and apprehended with a new and more specific meaning compared to the way people had commonly perceived them before. Darwin offered a new and more operational dimension to all these terms, a dimension often free of the animistic and teleological metaphors that had sometimes characterized the use of these terms in the past (see chap. 6).

Similarly, with the advent of Mendelian genetics, new definitions and terms emerged. For instance, it became particularly difficult to conceive of the behavior of genetic material in terms of the visible qualities of organisms as had been the case in the past when natural biologists spoke of "mixture." The term was associated with the appearance of "intermediate" characters, especially in cases of crossbreeding (e.g., the mating of a male donkey with a female horse giving birth to a mule). For de Vries, "the individual as a whole was . . . nothing but an enlarged version of the original set of pangenes in the nucleus of the fertilized egg (zygote)" (Mayr 1982, 781).

From this perspective, it was not necessary to specify whether the term *mutation* referred to the visible trait or to an underlying germplasm. For breeders and farmers, however, such an assimilation was unthinkable because they already knew from past experience that many of the traits for which they were breeding were influenced as much by environmental factors as by genetic factors. Observation required the decoupling of genetic factors from visible results at the trait level. In 1911 Wilhelm Johannsen was the first geneticist to stress the importance of distinguishing between different levels of observation. To this end, he defined the concept of "phenotype" and contrasted it with that of "genotype"—the phenotype was considered the visible level, whereas the genotype was the invisible level (Johannsen 1911). The formulation of a number of key concepts—including those of "genotype" and "phenotype"—but especially that of "gene" and the distinction between the general notion of "inheritance" and the precise principle of "heredity" based on Mendel's laws and their consequences, was crucial to the foundation of the nascent discipline of genetics. Of course the formulation of these concepts was also a source of debate. Biologists and theorists have continued to debate these issues ever since.

In this respect, discussions of the genotype-phenotype relationship (or "genotype-phenotype map," as it has been called, e.g., in Altenberg 1994) are part of a complex history that was accompanied by the formulation of the concept of "norm of reaction" (Woltereck 1909) in the early twentieth century, and later that of "phenotypic plasticity" in the 1960s (Bradshaw 1965). The last two concepts are closely linked as much as they are linked to the history of genetics and its integration into the biological sciences at the beginning of the twentieth century, the ins and outs of which I will now analyze.

Linking Plasticity to the Genotype

THE *REAKTIONSNORM* AS A DESCRIPTION OF INHERITED VARIABILITY

After the rediscovery of Mendel's laws and the emphasis on the discrete nature of Mendelian factors, the saltationist theory, which opposed to the Darwinian view of evolution, gained new popularity among zoologists. Richard Woltereck (1877–1944) belonged to the Darwinian camp. In June 1909, at a meeting of the German Society of Zoology to celebrate the centenary of Darwin's birth, Woltereck presented the results of several years' work based on his studies of different varieties of the crustacean *Daphnia*. The main purpose of his lecture was to offer a defense of the Darwinian view of evolution, showing that evolution occurs through natural selection acting on small continuous variations. Woltereck's study involved morphologically distinct pure lines (without genetic variation between individuals) of *Daphnia* and *Hyalodaph-nia* in different German lakes. These pure lines maintain their specific form through parthenogenesis over many successive generations. This meant that no genetic diversity could occur within the line (today we would say that they were almost like clones). Woltereck studied experimentally the influence of various environmental factors on a number of traits (phenotypes). Finally, he presented a very detailed account of the effects of food, temperature and density of conspecifics on the reproductive cycle of *Daphnia*, which alternated between a mode of parental reproduction and reproduction by parthenogenesis. Woltereck also moved some *Daphnia* from Denmark to Italy to see whether the environment (including temperature change) could alter some of their traits. Most of his results concerned continuous traits of the Daphnia (such as head height; see fig. 3.1) rather than discrete traits (as had been the case with Mendel's peas—yellow or green, flat or wrinkled). Woltereck

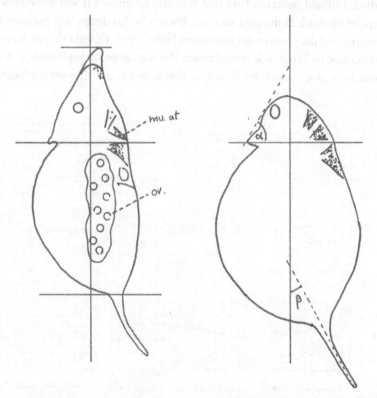

FIGURE 3.1. Lateral schematic views of females of *Hyalodaphnia culcullata* (*left*) and *Daphnia longispina* (*right*) from Woltereck (1909, 114, figs. 1, 2). In the left view the head height is measured along the vertical axis between the top horizontal line and the horizontal center line. The "relative height of the head" (which is the main feature of interest to Woltereck) is the height of the head divided by the distance between the lower horizontal line and the upper horizontal line (and multiplied by one hundred to be expressed as a percentage). (*mu. at*: muscles of the antenna; *ov.*: ovary). On the right, α and β are examples of other quantitative characteristics.

compared the traits together and in relation to the different levels of nutrients to which the crustaceans had access.

For the trait "relative head height," Woltereck found that the trait (1) varies between different pure lines, (2) is influenced by environmental factors such as nutrient levels, (3) varies almost independently of other factors such as temperature, and (4) has a cyclic variation that depends on other factors such as time of year (fig. 3.1). However, Woltereck also observed that the phenotypic response to a given environmental variation was not the same for different pure lines. He drew "phenotypic curves" to describe this phenomenon (fig. 3.2).

Phenotypic curves change with each new variable. There was therefore potentially an almost infinite number of them. Woltereck coined the term *Reaktionsnorm* to describe the set of relationships that link the curves together. According to him, it is the *Reaktionsnorm* that can be transmitted and thus inherited. I should point out here that Wilhelm Johannsen (I will come back to him and his work in the next section), known for his distinction between the genotype and the phenotype (Johannsen [1909] 1913), thought that each curve as described by Woltereck should instead be seen as possible phenotypic variations for a given genotype. Based on this analysis, the philosopher Sahotra

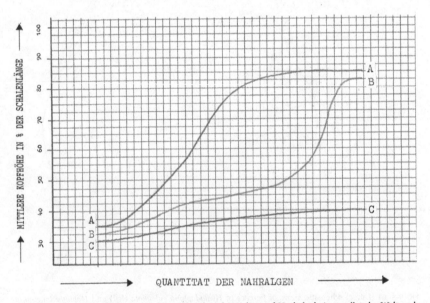

FIGURE 3.2. Phenotypic curves of three females of pure lines of *Hyalodaphnia cucullata* by Woltereck (1909, 139, fig. 12). X-axis: nutrient levels; y-axis: relative head height. A: *Moritzburg* line; B: *Brosdorf* line; C: *Kospuden* line. All lines were grown at an intermediate constant temperature and were from an "intermediate generation" of parthenogenesis. The curves show a nonuniform variation between pure lines.

Sarkar concluded that Woltereck interpreted the genotype through the reaction norm as a permissive agent of phenogenesis, whereas Johannsen saw it as a determinant (Sarkar 1999). More specifically, for Woltereck, all heritable variation occurs through the modification of the norm, which is not a single curve. It therefore seems almost impossible to predict the phenotype from the genotype. But the main result of Woltereck's work lay elsewhere. Indeed, since the *Reaktionsnorm* consisted of continuous variations in the phenotypic curves, Woltereck's work had theoretically shown that Darwinism was "saved" from saltationism because selection would act on small, gradual changes.

If Woltereck associated the *Reaktionsnorm* with the "genotype," he did so with a different conception than Johannsen. He explicitly stated that "the genotype of a quantitative trait . . . is the inherited reaction norm" (Woltereck 1909, 136, trans. Sarkar 1999, 236). Therefore, the result of the genotype was not the phenotype, but a process of *phenogenesis*. This word was mostly used by Soviet biologists to characterize a view they had in which the phenotype was seen as the result of genotype-environment interactions as opposed to a more deterministic view of the genotype. It was this latter interpretation that came to dominate in the West in the years that followed. Today, when biologists want to insist on genotype-phenotype interactions rather than on the result of genetic determinism, they speak of "process of differentiation." Woltereck's conception of the reaction norm found its greatest echo in the Soviet Union, where it became a central concept with both scientific and ideological implications.

Woltereck's 1909 *Reaktionsnorm* differed significantly from most contemporary definitions of "norm of reaction." Woltereck called "phenotypic curves" the functions representing the response of a genotype to a single environmental variable—what is now called norm of reaction—and he used the term *Reaktionsnorm* to characterize the totality of the curves that could potentially be drawn for a quantitative trait (Woltereck 1909, 135). Because it represented all the curves, and therefore all the potential genotypic responses to the environments, it was the *Reaktionsnorm* itself—that is, the totality of the curves—that could potentially be drawn for a transmitted quantitative trait. Woltereck thought of inherited mutations as the modification of the *Reaktionsnorm*.

While it is true that Woltereck's and Johannsen's theories of inheritance differed significantly in this regard—and I have also pointed out some of the differences they had with respect to Darwinism—it would be a mistake, in my view, to assume that the difficulties encountered with plasticity today are themselves related to alleged mistakes by Woltereck, as has sometimes been suggested elsewhere (e.g., Pigliucci 2001). Conversely, I would argue that

Woltereck's early twentieth-century conception of the relationship between hereditary and environmental factors is in many ways much more in line with contemporary research on plasticity (and genotype-phenotype interaction) than some of Johannsen's misinterpreted work. In the next section I will show the reasons why Johannsen's conception of the *Reaktionsnorm* was so strongly linked to genetic determinism and its rigid implications that dominated the early twentieth century.

It would be a mistake for any commentator to ignore the context of the time and to interpret Woltereck's definition of the *Reaktionsnorm* in light of our contemporary views. It should therefore be remembered that while posterity has retained Johannsen's work and writings as the initiator of the distinction and definitions for the terms *genotype* and *phenotype* (cf. chap. 3), during the first decades of the twentieth century the definitions of these terms remained unstable, and many authors, including Woltereck, sought to influence them. It was in this perspective that Woltereck, driven by his experiments on *Daphnia* clones, gave one of the sections of his paper the title "Genotypus = Reaktionsnorm" (Woltereck 1909, 135). He thought that the genotype (of a clone of *Daphina*), defined by the specific way it reacted to its environment, was not "determined" by any fixed value or factor (remember that Woltereck was writing more than forty years before the discovery of the structure of DNA by Watson and Crick in 1953).

Gerd Müller and Stuart Newman (2003), in their book *Origination of Organismal Form*, have shown that most contemporary debates are based on the meanings given to the words used by scientists rather than on the objects of study themselves—which brings us back to the old polemics. While the norm of reaction is now mostly understood as an evolutionary trait that could be subject to selection in itself (Schlichting and Pigliucci 1998; Nanjundiah 2003), its variability should not be confused with the variability of the trait that has not been stabilized by genes or their products (Müller and Newman 2003). Thus, the study of the variability of the norm of reaction must be associated with a trend that developed long after Johannsen, whereas the study of the variability of traits that have not yet been stabilized by genes and their products corresponds more to a trend that is directly linked to Woltereck's conception of the *Reaktionsnorm*. Thus, current concepts of plasticity still depend on the two divergent works of these authors, which in turn depend on the transformation of these concepts.

Many years later, and contrary to Woltereck's definition, the term *norm of reaction* would be defined as "each individual phenotypic curve" rather than "all the relationships described by these curves" (DeWitt and Scheiner 2004, 10–11).

DISTINGUISHING THE GENOTYPE FROM
THE PHENOTYPE

This brings us to a closer examination of Johannsen's work in its context. Johannsen's initial research was into the metabolism of dormancy, the metabolism of seeds and tubers, and bud germination in plants. He was soon able to show that dormancy could be broken with various anesthetic compounds, such as diethyl ether or chloroform. The identification of "pure lines" in the common bean *Phaseolus* is one of his most famous scientific contributions. He showed that seed size followed a normal distribution in the sense that it was subject to fluctuating variation even in a homozygous population for all traits, that is, without any genetic variation (such as clones obtained by parthenogenesis).

By focusing on mutation, Johannsen was in the mutationist camp and believed that evolution was a two-stage process: a first stage characterized by the occurrence of random mutations (which for mutationists corresponded to the creative part of the evolutionary process), and a second stage characterized by the maintenance or elimination of mutations by natural selection. Although Johannsen was convinced that this was how evolution worked, he proposed a distinction between "speculative evolutionary biology" (*Deszendenzlehre*) and the "study of heredity" (*Erblichkeitsforschung*). The philosopher Richard Burian has argued that "many pragmatically oriented geneticists utilized this restriction of the proper domain of genetics to argue against connecting genetics to such evolutionary speculations" (Burian 2000, 1130). Johannsen was one of them.

In his article "Om arvelighed i samfund og i rene linier" [About heredity in communities and in pure lines] (Johannsen 1903) and in his book *Arvelighedsloerens Elementer* [The elements of heredity] (Johannsen 1905), Johannsen introduced a number of new conceptual terms. In the last one—rewritten, expanded and translated into German under the title *Elemente der exakten Erblichkeitslehre* [The elements of exact heredity]—Johannsen introduced the term *gene* for the first time (Johannsen [1909] 1913, 124). This neologism replaced the common term *pangene*, which was based on "the Darwinian theory of pangenesis," which had acquired a strong theoretical connotation and was on the way to being completely rejected. Conversely, Johannsen thought,

[The term *gene*] should express only the simple conception that the characteristics of an organism are, or can be, caused or co-determined by "something" in the make-up [*Konstitution*] of the gametes. The word "gene" was completely free of any hypothesis. It expresses only the fact that the characteristics of the organism are determined in the make-up of the gametes by particular

"conditions," "factors," "units," or "elements." These are at least partially separable and thus to some extent independent—in short, exactly what we wish
to call genes. (Johannsen [1909] 1913, 143–44, translated in Burian 2000, 1130)

The formulation and definition of the notion of "gene" was a way for Johannsen to offer a term free of any evolutionary speculation. *The Elements
of Exact Heredity*, in which the neologism "gene" appears, is now considered
one of the canonical textbooks in genetics.

In December 1910 Johannsen, who was already famous in the international community of geneticists, was invited to give a speech to the American
Society of Naturalists. The text of the conference was published in March
1911 in the prestigious journal the *American Naturalist* (Johannsen 1911). In
the paper, he started with a recontextualization of the terms *heredity* and *inheritance*. Johannsen had borrowed these terms from colloquial vocabulary;
they first concerned the transmission of money, material possessions, ideas,
or knowledge. The idea of transmission between parents and offspring was
initially at the core of the debate. Johannsen stressed that such a conception
of inheritance was the most intuitive, albeit naive and old. The Darwinian
theory of pangenesis was not far away from such a conception. In many theories, whether in the Darwinian theory of pangenesis or in others (another
example is the inheritance of acquired characters of Lamarck), the idea began
to appear that the "personal qualities" of organisms were the inherited factors, or even, strictly speaking, their real "characters." Johannsen called this
the "transmission-conception of heredity," or the view of "*apparent* heredity"
(Johannsen 1911, 130). He thought that the supporters of eugenics could em-
ulate such a conception even though it added nothing to the biological problems of heredity. Worse, in his view, it suggested the opposite of what actually
happened. According to Johannsen, "The personal qualities of any individual
organism do not at all cause the qualities of its offspring; but the qualities
of both ancestor and descendant are in quite the same manner determined
by the nature of the 'sexual substances'—i.e., the gametes—from which they
have developed" (Johannsen 1911, 130).

Johannsen postulated that personal qualities were the "reactions of the
gametes" when they joined to form a zygote. Therefore, contrary to what the
transmission conception suggested, the nature of the gametes was not "directly" determined by the personal qualities of the ancestors. It is through this
idea of a "reaction of the gametes" that we can see how Johannsen's studies
in chemistry had influenced his analyses. Ultimately, he defined the modern concept of heredity, which he decided to call the "genotype conception
of heredity" in contrast to the old "transmission-conception of heredity." He

argued that the genotype conception of heredity was based on contemporary research in genetics and in particular the pure-line breeding research he had carried out as well as hybridization based on Mendel's model.

Johannsen realized that many of the terms used by geneticists at the beginning of the twentieth century belonged to an old trend and needed to be renewed. It was therefore time for Johannsen to reform the vocabulary of geneticists. Johannsen also realized that the discipline, which focused on heredity, was in a period of transition and on its way to becoming an exact science.

Johannsen proposed the terms *gene* and *genotype*, which laid the foundation for a new science of heredity based on genetics. He gave the following definitions:

> The "gene" is nothing but a very applicable little word, easily combined with others, and hence it may be useful as an expression for the "unit-factors," "elements" or "allelomorphs" in the gametes, demonstrated by modern Mendelian research. A "genotype" is the sum total of all the "genes" in a gamete or in a zygote. When a monohybrid is formed by cross fertilization, the "genotype" of the F_1-organism is heterozygotic in one single point and the "genotypes" of the two "genodifferent" gametes in question differ in one single point from each other. (Johannsen 1911, 132–33)

Aware of the limitations of the research in the emerging field of genetics at the time, Johannsen did not hypothesize the ultimate nature of these "genes." However, he was convinced that these genes could overlap with reality. Johannsen's recommendation was to use the noun *genotype* with caution, preferring the adjective *genotypic* and stating that "we do not know a 'genotype,' but we are able to demonstrate 'genotypical' differences or accordances." It is possible to account for these accordances and differences through breeding experiments, but it is not yet possible to identify genotypes directly.

Johannsen took a similar approach when he defined the notion of "phenotype." In order to pretend to explain the genotype, geneticists necessarily need an additional term because the only "visible" access to genotypes is based on the "qualities" or "reactions" of organisms. Some organisms with the same genotypic constitution may differ more or less in their personal or individual qualities when they develop under different environmental conditions. It is therefore sometimes difficult to know whether these organisms have the same "genotypic" constitution or not. Johannsen's solution is to say that these "qualities" are similar enough that they belong to the same "type." These identifiable "types" are what Johannsen proposed to call "phenotypes." Although they are more phenomenological in nature, phenotypes certainly correspond to real and identifiable objects.

With this definition of *phenotype*, Johannsen is then able to explain why the idea of evolution through continuous transitions, from a "type" to another, was one that emerged among biologists and botanists. It is precisely because the varying conditions of life are often the causes of fine phenotypic variation (often morphological)—what is mainly studied by biologists—that biologists have in mind the idea of a gradual evolution. But in Johannsen's view, these phenotypes should be linked to the genotypes. In this conception, it is truly the genotypes that are critical for evolution. Hence, this is the essential point of his distinction between the genotypic conception and the transmission conception of heredity; the latter conception was regarded as the so-called phenotypic conception of heredity.

But what was the evidence in favor of the view that the genotype conception of heredity was the one appropriate to reality? Johannsen based his argument on his observations of pure lines. He noted that one can always see considerable genotypic differences in "apparently" (phenotypically) homogeneous populations, that is, having only one "type" of individuals who differ only by minimal fluctuations. In fact the change of type in a population was not for Johannsen a result of evolution but of the genetic differences within the population. Genotype was not a function of a character or a personal trait of an ancestor. Johannsen compared the genotype of a gamete (or of a zygote) to a complicated physicochemical structure. In Johannsen's conception, a genotype reacts only in relation to its "realized state" and not according to the story at the origin of its creation. For this reason, Johannsen considered the "genotypic conception of heredity" as "ahistorical." For him, any "true heredity" relies on such an "ahistorical" conception similar to those found in chemistry or in any other exact science. Heredity may therefore be defined as "the presence of identical genes in ancestors and descendants."

The definition that Johannsen provided for the genotypic conception of heredity and his research on pure lines allowed him to settle Mendelism and make biology (specifically, genetics) a highly theoretical and scientific discipline. By focusing on stability, with the genotype conception of heredity (and where the transmission conception was mainly dealing with instability), he, at the same time, confronted biology with the principles of physics and chemistry. In the conclusion of his paper, Johannsen insisted on the fact that the genotypic conception of heredity did not provide a complete explanation of heredity. Instead, it should be seen, at best, as a tool to guide future research. He even suggested that this tool "may be proved to be insufficient, unilateral and even erroneous—as all working-hypotheses may some time show themselves to be." Despite the great success of Johannsen's paper, many uncertainties would remain regarding the final status of the definition he offered. This would be the case, for instance, with the notion of *Reaktionsnorm* that Johannsen had borrowed from Woltereck.

In the article of 1909, Woltereck had defined the *Reaktionsnorm* as "the sum total of the potentialities of the zygotes in question," and Johannsen had considered these potentialities to be partially separable from each other (through segregation). Consequently, Johannsen then considered the genotype to be perfectly capable of expressing the same idea, since the genotype was composed of genes, which could be defined in terms of "separate potentialities." Johannsen attempted to integrate Woltereck's more specific idea into his own definition of the genotype, namely, the idea that the organization should be seen as a whole despite the apparent diversity of the organism's "reactions." For Johannsen, it was the genotype that directly determined these multiple and variable "reactions" of the organism through its interactions with all the incident factors (whether internal or external). For this reason, according to Johannsen, the concept of *Reaktionsnorm* was consistent with the genotype conception of heredity as he defined it. This was true even though, by equating the concept of *Reaktionsnorm* with that of the concept of genotype, Johannsen tended to "smooth out" (without completely eliminating) the nuance Woltereck had suggested between "the variability of the reaction norm" (what Woltereck called *Reaktionsnorm*) and "the trait variability" (what Johannsen saw as the *Reaktionsnorm*). In other words, where Woltereck emphasized variability, Johannsen preferred to emphasize stability.

The formulation and the definition of the concepts of "gene," "genotype," "phenotype," and "*Reaktionsnorm*" or "norm of reaction" would not help to close the several debates concerning the organism-environment interaction. These debates covered both the question of heredity (specific to the emerging field of genetics) and the question of the origin of evolutionary change or variation (specific to the emerging field of evolutionary biology). Johannsen's influence would be significant for future work that focused on gene-environment interaction (the study of norm of reaction). It would also be critical in the neo-Darwinism context that developed in between the 1930s and 1940s, which attempted to summarize data for both genetics and evolution. I will now focus on these aspects.

THE GENOCENTRIC INTERPRETATION OF
THE NORM OF REACTION

Ultimately, Johannsen endorsed the concept of reaction's norm formulated by Woltereck. Moreover, he considered it to be "almost synonymous" with what he called "genotype." However, in Johannsen's opinion there was no

inconsistency between the fact that the phenotypic curves may vary considerably and the existence of constant genotypes. For him these curves represented "the *phenotypes* of the special characters, *i.e.*, the *reactions of the*
genotypical constituents" (Johannsen 1911, 145). According to Johannsen's interpretation of Woltereck's work, the specific contribution Woltereck made
to genetics was to provide a quantitative picture of what Nilsson-Ehle would
later call in 1914 *plasticity*, whereas Johannsen ascribed an adaptive meaning
to the concept of norm of reaction.

Herman Nilsson-Ehle (1873–1949) was a Swedish geneticist who carried
out experiments at the plant-breeding station in Svalöf beginning in 1900.
He is known for the resolution of one of the many apparent exceptions to the
Mendelian rules: namely, the inheritance of continuously varying or fluctuating characteristics (Müller-Wille 2005, 467). He demonstrated that

> By cross breeding two pure lines [homozygous combination, or the "genotype"
> of Johannsen], we can produce an amazing variety of new forms that can be
> made constant. A single cross breeding of this type is enough to produce the
> whole variety of forms or constant lines of an indigenous race. In other words,
> a whole "population" of new forms or lines is obtained by cross-breeding only
> two constant forms. (Nilsson-Ehle 1913, 862–63)

The article shows that in 1913 Nilsson-Ehle was fully aware of Johannsen's
genotype-phenotype distinction.[2] He, therefore, appears to be the first scientist,
and more importantly the first geneticist, to use the word *plasticity* to specifically describe the effect of the environment on the phenotype of an organism.
Nilsson-Ehle used the term *plasticity* [*plasticitet*] to describe the acclimatization of alpine plants to their environment (Nilsson-Ehle 1914, 543). He was also
the first biologist to describe the plant *Polygonum amphibium* as "particularly
plastic" because it could develop both terrestrial and aquatic characteristics according to the environmental signal it received. This plant was subsequently
then used as a model system in plasticity studies (Mitchell 1976). However,
Nilsson-Ehle's work has not influenced much of the contemporary literature
on plasticity because he didn't actually define the term *plasticity* precisely. Furthermore, he mainly understood this plasticity as a synonym for *purely adaptive*
(as opposed to *evolutionary adaptive*, more like *physiological adaptive*) and he
considered it as part of self-regulatory mechanisms (mechanisms by which an
organism responds to environmental changes). Therefore, Nilsson-Ehle, contrary to all expectations—since he accomplished his works in the context of
the emerging Mendelian genetics—did not define plasticity as "a property of
a single genotype," as Johannsen (1911) had understood the "norm of reaction"
as the "reaction range" of a single genotype to changing environments. The fact

that Nilsson-Ehle did not refer to "plasticity" in accordance with theories developed by Johannsen—one of the most influential scientists in the emerging discipline of genetics at the time—may explain why the notion of plasticity was not further disseminated before Bradshaw's seminal study in the mid-1960s.

In Britain and in the United States, where Johannsen's distinction between genotype and phenotype became one of the foundations of genetics, the causal consistency and efficacy of genotype would be emphasized in the following decades to the detriment of the complexity of genotype-environment interactions. The norm of reaction itself remained a relatively unknown concept during this period, with few researchers conducting studies in this area (DeWitt and Scheiner 2004, 10–11). At most, among the various interpretations that were offered, the norm of reaction tended to become a theoretical tool of genetics where previously it had been used in the context of understanding phenotype-environment interactions. With this in mind, one might ask to what extent the use and interpretation of the norm of reaction has influenced the further understanding of phenotype-environment interactions. Some authors, such as Massimo Pigliucci, seem to support the validity of such a reading of the norm of reaction, which should be related to phenotype-environment interactions rather than to the history of genetics. However, one may wonder whether Johannsen's conceptual modification of the notion of a norm of reaction by a certain "oversimplification" of the notion (i.e., by leaving aside the complexity of the description of the interactions) has contributed to a kind of conceptual gap of the idea. Let me now examine the conceptual transformation of the notion of norm of reaction in more detail.

After Johannsen latched on to the notion, the concept of norm of reaction became, first, a standard-bearer of the complexity of the nature-nurture interaction[3] (Hogben 1933) and, second, a conceptual tool for the analysis of phenogenesis[4] in general (mainly in the Soviet Union). In 1933 Hogben used data from Krafka, who in 1920 had published one of the first graphical representations of a norm of reaction (it represented the ratio between the number of facets in a *Drosophila* mutant and temperature; see fig. 3.3), to argue for an "interdependence of nature and nurture"—that is, of genetic and environmental factors—because the curves representing the norms of reaction were not parallel. According to Hogben, this observation illustrated the fact that phenotypic variation, which had already been measured by the variance (Fisher 1918), could not be decomposed additively into genotypic and environmental components because there remained a variable interaction between genotype and environment. This led Hogben to argue against a facile genetic reductionism, that is, the claim that phenogenesis can be entirely explained from a genotypic basis (Sarkar 1999, 239).

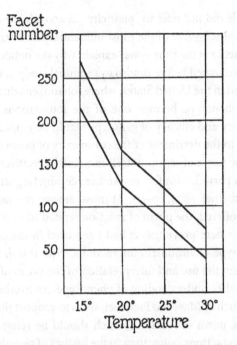

FIGURE 3.3. Norms of reaction for the dependence of the number of eye facets on temperature in males *(top)* and females *(bottom)* of *Drosophila melanogaster* (from Krafka 1920, 419, fig. 5). *Abscissa*: temperature; *ordinate*: average number of facets. The upper curve corresponds to the unselected stock of males; the lower curve corresponds to the unselected stock of females. Krafka drew two main conclusions from his observations on *Drosophila*: (1) the average facet number at a given temperature differs within stocks, and (2) the difference in the average number of facets between two temperatures is not a constant for all stocks. In other words, the number of facets is determined by a specific germinal constitution *plus* a specific environment.

But Hogben's argument would not be taken up by his successors because in the 1930s and 1940s, Haldane's algebraic analysis was the most influential method. In addition, during this period, "genetics and evolutionary biology coalesced around a Mendelian core" (Sarkar 1999; see also Sapp 1987). This involved attempts to (1) reticulate the structure of Mendelian genetics to encompass the complexity of phenogenesis, and (2) delineate exactly the genotypic contribution to phenogenesis and, in the spirit of (1), impute as much causal efficacy to the genotype as possible (Sarkar 1999, 239).

In fact as early as 1914 William Bateson had argued for the primacy of Mendelian analysis in exploring questions of inheritance for other disciplines of biology. At the same time, Bateson had pointed out that understanding the mechanisms of inheritance was central to a coherent theory of evolution: "Earlier it was hoped [by the ex-embryologists turned geneticists] that by a simple inspection of embryological processes the modes of heredity might be ascertained, the actual mechanism by which the offspring is formed from the body of the parent" (Bateson 1914, 289).

Because Bateson believed that the description of the modes of heredity had not been achieved, he questioned the efficacy of embryological methods as well as those of cytology. He also advocated a clear separation between the disciplines of genetics and embryology (Sapp 1987, 42). From this moment on, it became possible to think about heredity and about genetics without any reference to embryology, unlike what had been the case immediately after the rediscovery of Mendel's laws.

Sarkar has also described how at the same time (from the early 1920s) an active genetics research program emerged in the Soviet Union. It was concerned with the role of various factors (genetic and external) in phenogenesis. Some geneticists carried out experiments on *Drosophila* at the beginning of the program (Romaschoff 1925; Timoféeff-Ressowsky 1925; Timoféeff-Ressowsky and Timoféeff-Ressowsky 1926). Their results suggested that there was no clear trend in favor of the dominance of genetic or environmental factors in the determination of the phenotype. Instead, there remained a systematic uncertainty about the relationship between environment and phenotype.

In 1926, however, Oscar Vogt (1926) proposed a gene-centered interpretation of these results. He introduced two new concepts: the concept of "expressivity" to describe, in probabilistic terms, the extent to which a genetic mutation is manifested, and the concept of "penetrance" to describe the proportion of individuals carrying a genetic mutation who show any effect at all. With such definitions, Vogt neglected the differences between different lines—differences that Romashoff and Timoféeff-Ressowsky had noticed. However, Sarkar showed that since 1920, the two notions of "penetrance" and "expressivity," as biologists now call them, had been interpreted as properties of the mutation (and, eventually, the allele) rather than as properties of a mutation relative to a constant genetic background (which is what they actually are). These terms were also used by Conrad Hall Waddington with the same definition as Vogt, and they were introduced in the United Kingdom in 1938 (Waddington [1938] 2016). Waddington wrongly attributed the terms to Timoféeff-Ressowsky, and like Vogt, he neglected the differences between pure lines. This misinterpretation was not without consequences, as the complexity of genotype-environment interaction disappeared in Waddington's interpretation, in Britain in general, and even in the United States, since it was Waddington who introduced the notions into English-speaking countries. The modified concepts retained a strong genetic etiology, linked to the notion of norm of reaction, over the idea of a variability (plasticity) of the phenotype induced by the complexity of genotype-environment interactions.

Plasticity, expressed in the phenotypic manifestation of a trait, became the result of the expressivity of a gene and (indirectly) of its penetrance. Meanwhile, in the West (in the United States and in Europe, outside the Soviet Union), Ronald Fisher (1890–1962) in 1918 and Sewall Wright (1889–1988) in 1920 proposed methods almost identical to those just mentioned, which tended to undermine the complexity of genotype-environment interactions.

With the advent of statistical methods in the 1950s, broad heritability (individual heritability) was understood as the proportion of phenotypic variation due to genotypic variation. Therefore, genotypic-environmental interaction was considered identical for all genotypes in all environments, and penetrance was defined as the conditional probability (based on Bayes's theorem) that a given genotype expresses a given phenotype. Variability in the phenotypic expression of a trait became the result of gene "expression" and, indirectly, of its "penetrance." Environment was no longer an issue, and the distinction between "expressivity" and "penetrance" tended to disappear from the literature.

The Adaptive Interpretation of the Norm of Reaction

Back in the Soviet Union, the notion of norm of reaction was used from 1910 to overturn the principle of the inheritance of acquired characteristics (Blacher 1982). In 1926 Dobzhansky observed the effect of temperature on the color of the Chinese primrose *Primula sinensis*. He wanted to show that it was not the phenotype that was inherited from the parents but the norm of reaction. At room temperature, one strain of primrose produces white flowers, whereas at the highest temperatures, it produces red flowers (Blacher 1982, 132). Dobzhansky shows that even if the parent and offspring have flowers of different colors, the norm of reaction of the plants in response to a particular environmental signal (i.e., temperature) remains unchanged. In this way, the norm of reaction becomes, strictly speaking, constitutive of the unity of Mendelian inheritance. Soviet geneticists then distinguished between the adaptive norm of reaction and the nonadaptive norm of reaction. Where the adaptive norm of reaction is inherited, the nonadaptive norm of reaction is not. This distinction is based on models of organic selection. In 1896 the American philosopher and psychologist James Mark Baldwin (1861–1934) was the first to describe "organic selection" after conducting studies on children's learning (Baldwin 1896). Baldwin's idea, which differed from the Lamarckism of the time, was that behavior could influence and even facilitate the effect of natural selection. He showed that individuals could survive by adapting to adverse environmental conditions because of their abilities to learn.[5] If the environment does not change too abruptly, the most adaptive mutations will turn into congenital traits that were first learned. Consequently, learning "guides" evolution because it introduces a bias toward persistent mutations. The ability to learn increases the genetic variance of the population. During a sudden environmental change, only very different individuals (those that exist because of the ability to learn) can survive. Learning "accelerates" evolution and leads to evolutionary leaps. The paleontologist Henry Fairfield Osborn (known as a eugenicist) and the ethologist and psychologist Lloyd Morgan (1900) had specified the notion of "organic selection." What Baldwin called "organic selection" was first described by Osborn as a "mode of evolution requiring neither natural selection nor the inheritance of acquired character" (Osborn 1897, 949). The three authors had very different views. Osborn profoundly disagreed with the idea that "the power of plastic modification to new circumstances, or . . . 'self-adaptation,' [was] in itself a result of natural selection" (Osborn 1897, 950). In his view, natural selection was responsible for a "power of plastic modification" in organisms that was insensitive to new conditions and, in most cases, substantially adaptive. Osborn saw the "plastic modification" as an inherent power, constitutive of the protoplasm. Such a conception was rooted in the embryological and zoological tradition of Hans Driesch, Edmund Wilson, and Thomas Morgan. Osborn characterized his approach as follows:

> The only cases in which self-adaptation may be demonstrated as produced by natural selection are where organisms are restored to an environment which some of their ancestors experience. We can imagine that the adaptive response to the old environment is something which has never been lost as in the well-known reappearance of the pigment in flounders. (Osborn 1897, 950)

Around the same time, Baldwin from Princeton University and Lloyd Morgan, then professor of zoology and geology at the University of Bristol (later professor of psychology), came to the same conclusions. Both agreed that the term *variation* should be restricted to congenital variation and that the term *organic selection* should replace the term modification for all ontogenetic variation. In other words, the term *organic selection* makes it possible to reflect the process by which individual adaptations guide evolution. Unlike Osborn, the two authors did not see organic selection as an inherent property of the individual. In an article published in *Science* in 1906, the American zoologist Maynard M. Metcalf pointed out that "in their discussions of 'organic selection,' Morgan, Osborn, Baldwin and others have urged the importance of the plastic response of the individual members of a species in guiding the course of its evolution" (Metcalf 1906, 786). Although Metcalf agreed with all these authors on this last point—that organic selection exerts a decisive influence on evolution—he felt that the extent and precise nature of this influence was still under debate. Metcalf argued that there was another possible influence of plasticity that had not been properly considered. Metcalf relied on paleontological evidence that, from the beginning of the twentieth century, emphasized some trends in evolution. But Metcalf argued that these trends were linked to specific conditions in organisms rather than to environmental factors. He showed, based on his own experiments, that "the appearance, generation after generation of the same mutants of *Oenothera lamarckiana*, in numbers far greater than could be explained by purely fortuitous variation, is a further indication of some internal control over variation, making it somewhat determinate instead of purely indeterminate" (Metcalf 1906, 787). After the 1906 article, Metcalf thought that it would be possible in the near future to show evidence of "well-defined trends in evolution." This, he thought, could be done by extensive observations in breeding to see whether variations and mutations tended to cluster in certain directions rather than being equally distributed in all directions from the mean. It is worth noting here that Metcalf's advice would be followed by a growing number of biologists after the development of population genetics. As for plasticity, it was not until the mid-1960s that the term reappeared in genetics. But even if the word plasticity was less used in genetics before the mid-1960s, biologists were still referring to the notions of "norm of reaction," "plastic modification," "organic selection," and so on. These terms refer to "plastic" phenomena in life in a passive way—as the ability of an organism to produce a range of forms depending on the environment it encounters. It is after the debate concerning organic selection that the distinction between adaptive norm of reaction and nonadaptive norm of reaction appears.

The principle of "adaptive selection" was not elucidated until 1947, following the work of Georgii Gause (1910–1986), which was based on experiments with paramecia submitted to various environmental constraints. In a major article published after the Second World War titled "Problems of Evolution," Gause explained the reasons why the work of his predecessors had been forgotten. In Gause's view this was because when the debate about organic selection arose, the classification of variation into two distinct categories—modifications and mutations—was not well established, and the study of natural selection was still in its infancy. However, Gause showed that Efim Lukin, a professor of zoology, had independently rediscovered the principle of organic selection in the Soviet Union in 1936 (Lukin 1936). Gause described the four main principles that Lukin had arrived at concerning the relationships between phenotypic variability and genotypic variability:

(1) organisms frequently respond to environmental changes by adaptive phe-notypic modifications, (2) similar adaptive characters may be genotypically fixed in races normally living in the corresponding environments, (3) it is proven that conversion of modifications into mutations is not possible, and therefore (4) modifications can only be "substituted" by coincident mutations if the latter are associated with some advantages in the process of natural selection (Gause 1936, 22).

For Gause, organic selection was the assimilation of a modified phenotype through the integration of a genetic mutation into the organism's genome. The effect of the mutation was the same modified phenotype. Gause found that the less "genoadpatation" there was, the more adaptive changes were ob-served. Conversely, the more the adaptation was genetic, the less the pheno-typic changes were adaptive (Gause 1936, 65). The process described by Gause would then be called the Baldwin effect. Ivan Ivanovich Schmalhausen (1884–1963) was the first to popularize the process, replacing the notion of "organic selection" with "stabilizing selection" (Schmalheusen [1949] 1986). He empha-sized stabilization of an adaptive phenotypic response by its transformation into a genotypic response, which then leads to the same phenotypic results perpetuating its transmission through generations. Schmalhausen based his demonstration on mutagenesis experiments carried out by Dobzhansky in *Drosophila* in 1926 (Dobzhansky 1926). Dobzhansky had shown that a mu-tant *Drosophila* phenotype with an abnormality in abdominal development could remain silent for many generations if the food consumed by the ani-mals remained dry. However, the abnormality would reappear immediately if the offspring were fed a wet diet. From these observations, Dobzhansky concluded that if environmental factors could induce a particular trait, then the unaltered norm of reaction would be inherited. Schmalhausen referred to Dobzhansky's argument but made a distinction between "adaptive norm" and "morphose" (Schmalhausen [1949] 1986, 7–8). An "adaptive norm" is when the expression of adaptive change is such that it transforms the whole orga-nization. "Morphoses" are different: they arise as new and one-off responses without any historical basis (responses linked, e.g., to the impact of new en-vironmental inputs or to a modified norm of reaction due to a genetic muta-tion). This idea of the "adaptive norm" became essential and quickly attracted attention because it claimed that all modifications were only possible within the strict limits set by the norm. It was not the modification itself, but the "containment" of that modification within very precise limits that had to be seen as the adaptation. Schmalhausen thought that adaptive modifications were stabilized by "self-regulation." In 1942 Waddington in England proposed a similar idea, that of "canalization" (Waddington 1942). He distinguished between "genetic canalization[6]"—the ability of a genotype to produce two different phenotypes—and "environmental canalization"—the ability of a given genotype to produce a unique phenotype in different environments. Overall, canalization should be understood as the "intrinsic robustness" of developmental processes in response to genetic or environmental perturbations. At the same time, Waddington also defined the notion of "genetic assimilation," which was linked to the "Baldwin effect." Genetic assimilation implied that the new canalized trait would eventually be genetically stabilized whether the environmental circumstances at the origin of the perturbation remained the same or not (see chap. 7). As Sarkar has pointed out, when Dobzhansky brought the notion of the "adaptive norm" to the West, and more precisely to the United States, he insisted on showing that a mutation does not change a particular morphological character but instead introduces a change in the norm of reaction (Sarkar 1999). The notion of "reaction range" even frequently replaced the notion "norm of reaction" in some textbooks (Sinnott, Dunn, and Dobzhansky [1925]) 1950, 22). The authors of these textbooks then argued that phenotypic variability in the reaction range was a way of showing how phenogenesis depended on heredity-environment interactions and a way of emphasizing "environmental plasticity." Dobzhansky's use of the notion of norm of reaction led to its conceptual modification, as from 1955 on it became mainly a problem of population genetics because the norm of reaction itself was interpreted as a unit of Mendelian inheritance.

Dobzhansky's interpretation of the norm of reaction introduced a difference between his and Schmalhausen's conception. In a sense Dobzhansky followed Metcalf's suggestions when he insisted on the importance of focusing on the determined tendencies of variation rather than on its indetermined tendencies (Dobzhansky 1955, 786). On the other hand, Schmalhausen's problem mainly concerned phenogenesis, which was the main topic among Soviet biologists at the time. Dobzhansky, for his part, offered a kind of compromise between the Soviet view of the norm of reaction and a more genocentric view that had first been initiated by Johannsen.

The main point is this: from the moment the norm of reaction was seen as an adaptation, like any other trait, it became a potential target for natural selection (see chap. 5).

A Brief Interlude with the "Norm of Reactivity"

In the complex history of plasticity in genetics, the work of Richard Goldschmidt (1978–1958) occupies a special—and from a theoretical point of view, somewhat unusual—position.[7] Goldschmidt was an Austrian geneticist and embryologist who in 1940 published *The Material Basis of Evolution*, a book that summarized his main work. It introduced his famous—but highly controversial—theory of the "Hopeful Monster." Although much of his work won him the admiration of his peers (e.g., his work on sex determination), his rejection of the classical view of the gene and his idiosyncratic views on evolution (most notably his theory of the Hopeful Monster) damaged his scientific reputation in the 1940s. Goldschmidt was neither like Wright or Dobzhansky, who integrated mathematical formalism into population genetics, nor like Lysenko, who strongly rejected the foundations of Mendelian genetics.[8] He was, however, firmly opposed to neo-Darwinism, which sought to integrate population genetics, the Mendelian theory of genetic transmission between generations, and the Darwinian theory of natural selection. Goldschmidt (like Johannsen before him) was a true proponent of the "saltationist" theory of evolution. Although he was well aware of the work of Fisher, Haldane, and Wright in population genetics, he warned his readers throughout his seminal book against this kind of approach. In this regard, Goldschmidt reiterated the recommendation of his predecessor Johannsen concerning Galtonian biometry: "It is necessary to remember an old remark of Johannsen in his criticism of Galtonian biometry; namely, that biology must be studied *with* mathematics but not *as* mathematics" (Goldschmidt 1940, 137). This means that for Goldschmidt, any mathematical treatment, however brilliant, was useless if the biological evaluation of the material was wrong. Goldschmidt thought that this criticism, first formulated by Johannsen in relation to genetics, also applied to the mathematical study of evolution.

From this moment on, Goldschmidt's work took on an air of originality compared to most of the other work being done in the field at the time. Unlike Bateson, who questioned the efficacity of embryological methods in assessing the modalities of inheritance, Goldschmidt thought that a synthesis between genetics, embryology, and evolution was not only possible but necessary.

Goldschmidt's "genetical and developmental potentialities" corresponded to the useful "hereditary differences," since "the material basis of evolution" (which became the title of his book) was, in his opinion, the genetical and developmental potentialities of the organism (Goldschmidt 1940, 3). For Goldschmidt, all hereditary differences in the organism also implied a definite change in the development of the organism. The possibility and rate of genetic change depended on the rate of possible modifications during development. (Such modifications occur without counteracting integration and the "proper" functioning of embryonic processes.) Goldschmidt (1940, 6) concluded that the "potentialities of individual development are among the decisive factors for hereditary change and therefore for evolution." Therefore, in his opinion, experimental embryology cannot and should not be neglected in the analysis of evolutionary processes (for the reasons given above). By focusing on "developmental potentialities," Goldschmidt incorporated "active" implications into the concept of plasticity in genetics, implications that were to be found in Spemann and Mangold's embryology at the beginning of the twentieth century.

In the section of his book titled "Evolution and the Potentialities of Development," Goldschmidt focuses on the "norm of reactivity" and considers it as a tool for understanding evolutionary phenomena at the macroscopic level. "Norm of reactivity" is Goldschmidt's translation of Woltereck's *Reaktionsnorm*. He distinguished the notion of "norm of reactivity" from Johannsen's notion of "norm of reaction." He agreed with Johannsen on the definition of the "norm of reaction" (Goldschmidt 1940, 259). However, Goldschmidt thought that there was a difference between the norm of reaction as described by Johannsen and the *Reaktionsnorm* of Woltereck. The consideration of such a distinction is unique in the field because most authors after Johannsen no longer referred to Woltereck's subtleties, which were underpinned by the notion of *Reaktionsnorm*. Instead, they all seemed to adhere to a unique definition of the norm of reaction that followed Johannsen's genotype-phenotype distinction. Goldschmidt is the only geneticist to question such a distinction, since "the genotype cannot be described simply in terms of the phenotype." He criticized Johannsen's description of the phenotype because he thought "the description must contain the whole range of reactivity of the phenotype under different external or internal conditions" (Goldschmidt 1940, 252).

For Goldschmidt, the only way to reflect on what the "norm of reactivity" is and what its limits are is to analyze a wide range of facts linked to the potentialities of development. He supported such a statement by referring to a number of particular biological facts and examples. Among these examples, some recurred frequently, such as the seasonal dimorphism of butterflies, the sexual dimorphism of bees, and the dimorphism of the shape of leafy plants depending on their aquatic or terrestrial environment. Thus, he concluded that under different determined conditions (external or internal), development could easily be modified either slightly or greatly. Consequently, any major change could lead to significant changes in categories (e.g., from one species to another).

Goldschmidt drew evolutionary conclusions from his observations of "the potentialities of development." He thought that developmental potentialities were themselves "embodied in the potentiality for shifting processes of embryonic determination (including mutations), as providing a potential range of phenotypic changes of the order of magnitude of macroevolutionary changes" (Goldschmidt 1940, 265–66). Therefore, simple genetic changes involved in the developmental process, which can be conceived of as occurring in a single step, can result in very different individuals from the reaction norm of the species, possibly better adapted under certain conditions. Such a conception remained tied to a "saltationist" view of evolution and corresponded to his theory of "hopeful monsters." Goldschmidt argued that "the idea [is] expressed in the somewhat unconventional but plastic term 'hopeful monster'" (Goldschmidt 1940, 391). Apart from this (accidental?) occurrence, the word *plastic* does not appear elsewhere in the text. However, this occurrence is relevant because Goldschmidt was the only geneticist after Woltereck to refer to an active way of thinking about "plasticity" by introducing a distinction between "norm of reactivity" and "norm of reaction." However, the phrase "norm of reaction" always expressed, somewhat paradoxically for Goldschmidt, as for other authors after him, the passive meaning of the concept of plasticity—the ability to adopt multiple phenotypes—whereas the "norm of reactivity" referred to an active meaning of the concept—the potentiality/power to generate something new.

Goldschmidt's examination of the "norm of reactivity" provides an interesting interlude in the history of the theories of plasticity in genetics. In fact, because of his controversial position—he did not fit into the mainstream, which aimed to reduce the nuisance of the environment and to give a full account of the stability of certain biological phenomena—Goldschmidt focused on the "instability" of these biological phenomena and the active capacity of "reactivity" of living beings through their norm of reactivity. Furthermore, after Goldschmidt, the norm of reaction was further discussed, the focus shifting to its potential "stability," since the notion was understood mainly as an adaptation explained by its genetic basis. It was not until the mid-1960s that more attention was given to trying to understand what determines the norm of reaction, when Anthony Bradshaw elaborated the notion of "phenotypic plasticity" (see chap. 5).

4

Plasticity in Evolutionary Biology
A Boundary Concept

The relations of evolutionary plasticity to genetic plasticity encourage the hope that
progress in knowledge of the latter, which is accessible to experimental investigation,
will increase our knowledge of the former. And in both cases the variability which plasticity denotes will have to be studied from the embryological aspect.

GAVIN DE BEER (1940, 97)

Until the end of the nineteenth century, the term *plasticity* (and especially its
adjectival form) was used mainly in embryology (see chap. 2). In genetics its
use spread with the development of Darwinian evolutionary biology. Yet the
concept of plasticity differs in the study of evolutionary mechanisms and in
that of genetics (see chap. 3). Geneticists refused to adopt Darwin's use of the
term (when Darwin spoke of "plasticity of organisation"), and Darwin never
referred to plastic- terms in his work in the way that embryologists had done
before him.

The leitmotif of this chapter is to explore how plastic- terms have been
used in the field of evolutionary studies and how this field has influenced
embryologists and geneticists' later ideas about plasticity. To this end, I begin
with Darwin's use of the term *plasticity* in his work, with particular emphasis
on his seminal work *On the Origin of Species by Means of Natural Selection,
or the Preservation of Favoured Races in the Struggle for Life* (Darwin 1859), in
which he set out his theory of evolution by natural selection.

Although *plasticity* does not have the central and foundational place in
the Darwinian corpus that *variation* (see Winther 2000), *heredity*, or *adaptation* have, the term *plasticity* appears several times in Darwin's work. On
the other hand, Darwin used the term often enough for historians of science,
such as John Beatty, to say that the term *plasticity* was used as *synonym* for
that of *variation* (Beatty 1982, [1986] 1992, 2006). While the central concept
of "variation" overlaps with plastic- terms in Darwin's texts, the latter retains
the much narrower, technical meaning it had among breeders. Indeed, the
index to later editions of *On the Origin of Species* defined it as a disposition in
organisms: "plastic: readily capable of change" (Darwin 1876, 438). Therefore,
plasticity and related terms deserve special attention in Darwin's corpus.

This chapter is divided into four main sections. In the first, I examine how Darwin understood plasticity, arguing that he had two main uses of the term: plasticity as what would eventually be explained by natural selection—what he called "the plasticity of organisation"—and plasticity as a source of variations. Although these two uses of plasticity differ from the active use—the potentiality to generate or to induce a development—and the passive use—the ability to adopt at least two different forms (discussed in earlier chapters), I will draw analytical links between these meanings in this section. I then examine the alternative uses of the term *plasticity* in the evolutionary thought of Darwin's contemporaries. In the second section, I highlight the connection between plastic- terms and inheritance terms in the work of the French neo-Lamarckians and show how they introduced a confusion about the role of plasticity in evolution that persists in the life sciences. In the third section, I consider the forgotten work of Sir Gavin de Beer, the first theorist to attempt to provide a synthetic view of evolutionary processes by bringing together data from genetics and embryology. I discuss how de Beer uses plasticity as a boundary concept that can bridge the gap between different areas of life science. I revisit his idea in light of the distinctions between plasticity in embryology and genetics made in chapters 1 and 2. In the fourth section I focus on the modern synthesis of the 1950s and 1960s, the theoretical current that dominated evolutionary thinking after Darwin and that linked genetics to evolution. I show how the modern synthesis established a link between genetic diversity and evolution that led to a confusion very similar to that of the neo-Lamarckians, conflating the concept of plasticity in Darwin's theory of evolution by natural selection with that of inheritance (genetics).

Darwin's "Plasticity of Organisation" and Plasticity Understood as a Source of Variation

At the end of the nineteenth century, the majority of life scientists believed that some "external" factors controlled the course of evolution, such as the preadaptation of variation, divine intervention, or a spontaneous tendency of organisms to evolve toward greater complexity. Charles Darwin's theory of evolution by natural selection (Darwin 1859) highlighted a new aspect of the modeling of living things that, like the work of the Cambridge Platonists of the seventeenth century, didn't rely on a purely mechanistic understanding. Far from any form of seventeenth-century panpsychism, however, Darwin offered the life sciences a theory that would soon acquire the status of a science while preserving some autonomy from the physical and chemical sciences.

To build his theory, Darwin drew heavily on the work of breeders. From chapter 1 of *On the Origin of Species*, Darwin noted that it was possible to observe an infinite number of structures and constitutions in nature, with varieties and subvarieties differing very slightly from one another. This idea and the terms he used were borrowed directly from breeders. He goes on to claim that "the whole organisation seems to have become plastic, and departs in a slight degree from that of the parental type" (Darwin 1859, 9). Darwin himself later testified the use of this idea by breeders: "Breeders habitually speak of an animal's organisation as something plastic, which they can model almost as they please." This is reminiscent of the active sense of plastic- terms used by early philosophers. In the sixth edition of *On the Origin of Species*, he finally defined "plastic" in this sense as something "readily capable of change" (Darwin 1876, 438), a definition that remained vague. However, Darwin explicitly claimed that he derived the use of the plastic- term not from any theoretical usage but from the vernacular use of the term by breeders.

However, the "plastic" nature, which is manifested by a certain "tendency to variability" observed by the common sense of the breeder, is not the work of man. The breeder merely preserves or encourages the accumulation of variations that can occur after organisms have been exposed to certain conditions. In other words breeders allow the preservation of some variations that would have disappeared in nature. In chapter 5, "Laws of Variation," Darwin takes a closer look at the question of character variability and its causes. In particular, Darwin is interested in understanding the reasons for the "tendency to variability" that he sees both in the breeders' species and in nature.

Darwin points out that some organisms of the same species can take on different characteristics depending on their living conditions. However, he does not establish a law of nature on the basis of these observations. Unlike Lamarck, who thought that environmental conditions were a determining factor in variation (e.g., the stretched length of the giraffe's neck from craning for leaves) (Lamarck 1809), Darwin does not consider the direct action of environmental conditions to be a determining factor in variation. Rather, he establishes a causal hierarchy between "living conditions" as the cause of variability (which he places at the bottom of the hierarchy) and the "nature of the organism" as another likely cause of variability (which he places higher in the hierarchy of causes of biological variability). While it remains difficult for Darwin, and for us, to understand what the specifics of this "nature of the organism" are, Darwin clearly thinks that it is precisely because the nature of the organism is plastic that it has a tendency to variability—a tendency that can be seen mainly in the variation of conditions.

Darwin uses a variety of examples of variation due to differences in conditions to illustrate his concept: shells that vary in color according to the depth at which they are found, the shape of the leaves of plants that can vary according to their climatic environment (aquatic or terrestrial), and so on. In his view, however, it would be possible to observe just as many similar variations in organisms in the absence of changing conditions. At this precise moment in his explanation, Darwin sees "variation" as a result or rather as an existing situation explained by variability as a cause, a disposition of living organisms to express different forms.

Consequently, for Darwin, variation is primarily of unknown causes, and variation occurs independently of environmental conditions. There is an inherent tendency in all species to produce variants. Therefore, it seems that the laws of variation should be sought in the "nature of the organism" rather than in "external conditions" because it is in the nature of the organism that one can hope to identify the causes of its variability. Darwin details this idea:

> The direct action of changed conditions leads to definite or indefinite results. In the latter case the organisation seems to become plastic, and we have much fluctuating variability. In the former case the nature of the organism is such that it yields readily, when subjected to certain conditions, and all, or nearly all the individuals become modified in the same way. It is very difficult to decide how far changed conditions, such as of climate, food, etc., have acted in a definite manner. (Darwin 1859, 106)

By "definite results," Darwin means that all individuals in a species respond in a similar way (including not responding at all) to a given set of environmental conditions. With indefinite results, individuals of the same species respond differently to a given set of environmental conditions. "Plasticity" in Darwin's terms can therefore only be seen in the case of "indefinite results," that is, cases where individuals of the species respond differently to similar environmental changes.

However, while Darwin was able to make such a distinction between the definite and indefinite results of external conditions, he did not identify the reason for the "plasticity of organisation" and confessed that he could not explain the cause of such phenomena: "Such considerations as these incline me to lay less weight on the direct action of the surrounding conditions, than on *a tendency to vary*, due to causes of which we are quite ignorant" (Darwin 1859, 107, my emphasis). Darwin's emphasis on a general, as yet unknown, intrinsic disposition to vary, not directed by the environment, is a clear example of how Darwin also used plastic- terms to bridge an epistemic gap. In considering the effects of the environment on variation, Darwin rejects

Lamarckism: he clearly shows that environmental change does not cause variability. Although we are ignorant of the tendency of individuals to vary, Darwin does not mean that it will always be impossible to identify the cause of variability. Here Darwin is occasionally interested (this will not be the core of his theory) in the *proximate causes of variation* without being able to iden-tify them one by one. His thoughts and intuitions on the subject of variability and its manifestations would be continued and deepened after him, for example by Richard Woltereck in his work on the *Reaktionsnorm*[1] (see chap. 3, sec. 3.2) and also later by the work of some breeders and genecologists[2] such as Ernest Babcock and Ray Clauser, who would offer classifications of different types of variation (Babcock and Clauser 1918). Most of these authors would support the Darwinian gradualist position (the idea that evolution proceeds by gradual small changes) in the face of a saltationist conception of evolution (evolution by large abrupt changes) defended by a growing number of zool-ogists (i.e., mutationists).

In the preface to the third edition of *On the Origin of Species*, published in 1861, Darwin returns to some of the ideas about the origin of species put forward by his predecessors and contemporaries. He argues that some of their ideas gradually enabled him to forge his own theory. He also points out that by the time he wrote *On the Origin of Species*, many eminent people, beginning with the French botanist Charles Victor Naudin (1815–1899), were convinced that "species, when nascent, were more plastic than at present" (Darwin 1861, xix). Darwin notes that Naudin, for example, based this con-viction on a "principle of finality":

> A mysterious, indeterminate power; fate for some; for others, providential will, whose incessant action on living beings determines, in all the epochs of the world's existence, the form, volume, and duration of each one of them, because of its destiny in the order of things of which it is a part. It is this power which harmonizes each member to the whole by appropriating it to the function it must fulfill in the general organism of nature, a function which is for it its *raison d'être*. (Naudin's text of 1852, quoted in Darwin 1861, xix–xx; in French in Darwin's text, my translation)

This preface shows the theoretical context to which Darwin refers when he uses the adjective *plastic*. Although his theory of the origin of species dif-fers from those of his predecessors, Darwin is in line with others for whom "plasticity" was a tool to refer to a notion at the limit of scientific explanation, a notion that remains specific to living things without referring to any mysti-cal power. For Darwin, too, the reference to the notion of "plasticity" serves to explain the limits of a theoretical argument and, in particular, to delineate what the theory of natural selection cannot explain: the proximate cause of variation and the tendency toward variability. Somehow it seems that what makes boundary concepts scientific (and powers/forces perhaps less so) is that they specify (point to) what the theory does not explain with the aim of eventually finding an explanation, whereas forces/powers fill the epistemic gap without explanation.

In both *On the Origin of Species* and *The Variation of Animals and Plants under Domestication*, Darwin uses the adjective *plastic* essentially in an "active" sense, similar to that used in embryology, as a disposition. However, Darwin dissociated this active use from all ancient architectonic views. When Darwin uses the adjective *plastic* or the noun *plasticity*, it is never in reference to any idea of "force" or "power." For Darwin, natural selection, far from being Naudin's "mysterious power," does not (as in Naudin's case) constitute a justification for the progressive loss of plasticity of species in the course of the history of evolution but instead tends to explain—as the ultimate cause of—their progressive modification over time. Darwin claims that he "can see no limit to this power [of natural selection], in slowly and beautifully adapting each form to the most complex relations of life" (Darwin 1861, 412). Natural selection is the explanatory basis for the progressive diversity of species over time. If diversity is due to natural selection, then change as a process depends on plasticity.

For Darwin, however, the term *plasticity* serves at most to fill in the explanatory limits of the theory when it comes to explaining the proximate causes of variation (the explanans of variation). Therefore, with Darwin, the concept of plasticity clearly appears for the first time as a semiempty conceptual space to be filled by future scientists working within this paradigm. Whereas in the past the concept of plasticity seemed useful to naturalists thanks to its vagueness—which did not distinguish between its passive and active nuances and thus could do a lot of "work"—with Darwin the concept of plasticity takes on a new dimension: it becomes a borderline concept, bringing together acquired scientific ideas with ideas in process.

Over the years Darwin found himself in the position of having to defend his theory, especially that which was most criticized, namely the status of variation. In this sense, the reference to *plasticity* would also be an attempt by Darwin to clarify what he meant by *variation*.

By drawing a link between the two terms—using one to underline the limits of the other—Darwin also introduced the possibility of confusion between the two, which tended to conflate their respective meanings for future interpretations. And indeed, some of his successors (and commentators) often confuse the two terms, treating them as synonymous (e.g., the historian Peter Bowler has argued that Darwin used *plasticity* as a synonym for *variation*). They conflate the use of "plasticity" as a disposition, or "what explains variation," and "variation" as "what is explained by natural selection." But let me explain what is surely at the root of the confusion in Darwin's writings.

As Jean Gayon has shown, Darwin is constantly challenged to explain how he understands the relationship between variation and selection in evolutionary mechanisms. And indeed, variation sometimes appears as an object of selection and sometimes as a result (Gayon 1998, 97). This is what Gayon called the "dilemma in the theory of natural selection" and with which Fleeming Jenkin (1833–1885) had challenged Darwin. In this view variation can either be seen as the result of the process of natural selection or as the object of selection acting on the variation itself, which is not really obvious because it is continuous. In the discrete case, the obviousness of the variety of forms is something to be explained. In the continuous case, the variation is an intrinsic disposition, but natural selection has acted on it to change the statistical appearance of the plastic trait in the population.

In other words Darwin needed to clarify whether it was necessary to explain the cause of variability to understand the selection process or whether, as he had repeatedly stated, selection acted on individual variations "indefinite" in direction. However, Gayon pointed out Darwin's apparent ambiguity in the fifth edition of the *On the Origin of Species* when he said that "the *efficacy* of the selective process may depend on the hypothesis concerning the nature of inherited variation" (Gayon 1998, 102, my emphasis). The underlying idea was that if we had a clear idea of the nature of variation, we might have a clearer idea of the selective process itself. Darwin agreed that explaining the nature of variation would help explain the nature of selection, although we do not need to know the nature of variation to understand that selection is at work.

This ambiguity is reflected in Darwin's use of the term *plasticity* to explain variation in later editions. Because the term can be defined in two different ways, reflecting both the cause—a tendency to variability—and the effect of variation—the result of the nature of the organism—it allows Darwin to maintain the ambiguity while implying a clarification. Nevertheless, the fact that he seems to lean in favor of using plasticity as a disposition or explanans (as evidenced by the definition of *plastic* he gives in the glossary of the last edition) is an indicator that, for Darwin, understanding variability is definitely useful for understanding the selection process even if it is not necessary.

The French Neo-Lamarckians:
Linking Plasticity and Heredity

As we have seen, Darwin's use of the term *plasticity* seems clearly linked to the issue of variation. Because Darwin showed that his theory of evolution had to account for both the variation of characters and the transmission of those characters, that is, the way in which variation is inherited, the question of variation soon appeared to be closely linked to the question of inheritance (Lewontin 1978). Because plasticity is linked to the question of variation by transitivity, it is also, in Darwin's case, intimately linked to the question of inheritance. This potential link between "inheritance" (transmission) and "plasticity" is not analyzed by Darwin. His provisional hypothesis on heredity, the "pangenesis theory"—which suggested that each part of the body could emit gemmules (small organic particles) that accumulated in the gonads to be transmitted from one generation to the next—was considered inadequate by his contemporaries. But if the relationship between these two concepts— inheritance and plasticity—remained unthought of by Darwin, it would be the subject of reflection by other evolutionary theorists, such as the French neo-Lamarckian transformists.[3] Among these authors, both the notions of "plasticity" and of "heredity" will form the fertile ground of an evolutionary current. We can already see that this tendency was quickly marginalized in the 1920s as the disciples of Mendel and evolutionary genetics came closer together. I argue, however, that neo-Lamarckians would still be influential in evolutionary biology through their use of plasticity.

Between 1880 and 1900, on the eve of the rediscovery of Mendel's laws, the idea of plasticity was omnipresent in the thinking of the French neo-Lamarckian transformists. The neo-Lamarckian transformist current first appeared in the United States at the end of the nineteenth century (the entomologist Alpheus Packard proposed the term *neo-Lamarckism* in 1885 to describe the views of the American transformist school). In his book *Qu'est-ce que le néolamarckisme?* (What is neo-Lamarckism?), the historian Laurent Loison defines neo-Lamarckism as follows:

> Because it is a causalist view of nature, in the classical sense of Cartesian mechanics, neo-Lamarckism, like Lamarck's theory, regards variation as an effect. Scientific explanation must therefore focus primarily on understanding its cause. Since evolution is conceived [by French neo-Lamarckians] as the algebraic sum of individual variations, the discovery of the cause of these variations fully explains the transformation of species. All neo-Lamarckisms are based on this conception. (Loison 2010, 10, my translation)

It would seem that in this view, in neo-Lamarckism variation is an effect of evolution. But as Loison notes, neo-Lamarckism holds that a full understanding of species transformation and evolution requires a full understanding of the causes of variation. Since this definition applies to all neo-Lamarckisms, the notion of plasticity, which is also widely used in their writings (Loison has shown that it was widely adopted by French neo-Lamarckians) and is often linked to that of heredity, will acquire much importance.

At the end of the nineteenth century, the issue of heredity began to raise many questions among biologists, particularly about the resemblance between parents and children. How could such similarity be explained? The French neo-Lamarckians did not believe that there could be any particular force at work in living bodies responsible for their evolution (Loison 2010, 12), something that had been defended by the American school, notably Louis Agassiz (1807–1873) and Alpheus Hyatt (1838–1902). Both adhered to an orthogenetic view of evolution (see Pfeifer 1965), the idea that organisms have an innate tendency to evolve in a particular direction (telos) due to driving forces. Therefore, they thought that the internal cause of variation was hierarchically superior to its external causes (i.e., environmental conditions). The French neo-Lamarckians refused to accept the idea of the existence of a specific internal force in living bodies responsible for their evolution. On the contrary, the French neo-Lamarckians believed that the tendency to evolve in a particular direction was mainly due to the inheritance of acquired characters, which they placed at the center of their view.

Such heredity has mostly been seen as a process of acquisition of variation at the individual level, thanks to environmental factors. Therefore, Loison describes—from a hierarchical point of view—that "the important phenomenon is first the capacity to acquire new characters at the level of the individual organism or the protoplasm (the basic cellular substance), and then, but only secondarily, to transmit them by heredity" (Loison 2010, 14). Heredity appears as a two-step process: acquisition and transmission. Based on this idea, the French neo-Lamarckist theory is based on two criteria: plasticity, which promotes the acquisition of new characters, and heredity, strictly speaking, which promotes the transmission of these characters. The underlying idea is that there is a disposition to change form in response to environmental conditions (plasticity) and that it isn't plasticity itself that is inherited but rather the specific form that the plastic body takes. From these two criteria of inheritance, it is therefore plasticity that allows us to explain one of the most important phenomena: the possibility of acquiring new characters. And indeed, Loison shows that between the 1880s and 1900s, "the idea of plasticity implicitly underpinned almost all of the experimental research of neo-Lamarckism" (Loison 2010, 33).

Moreover, because the French neo-Lamarckians would lose interest in evolutionary causal processes (i.e., what mechanisms cause changes) and focus almost exclusively on elucidating physico-chemical processes (on the part of organisms), they never produced a real explanation of evolution. The proximate cause of evolutionary variation would be reduced to its simple environmental determinism.

The confusion introduced by the neo-Lamarckians into evolutionary thinking about plasticity is therefore based, in my opinion, on two things that are closely related. On the one hand, it is based on an abusive assimilation between the phylogenetic notion of "adaptation" and the physiological notion of "habituation" (in other words, between the uses of the term *plasticity* in the field of evolutionary biology and those in the field of physiology). On the other hand, it is based on their semantic grouping of the proximate causes (explanans) and the ultimate causes (explanandum) of evolution. I will argue that this confusion persists to some extent even among those who are not modern proponents of Lamarckism (see chap. 7).

It is also important to note that in the work of the neo-Lamarckians, plasticity is sometimes understood as a core concept intended to explain the process of modification that precedes (or is included in) heredity in the pattern of evolutionary species transformation but also sometimes as "passive individual adjustment to environmental conditions." This use of the term is quite different from Darwin's. Darwin used the term *plasticity* to clarify and specify the cause of variation in the evolutionary process and never combined the two meanings he had of plasticity in the same explanation (although he might have used both). He showed that the term *plasticity*, unlike that of *variation*, allowed one to focus on the proximate causes of the phenomena of variation. However, Darwin's use of the term, by bringing it closer to the notion of variation, ultimately also leads to another confusion (between the proximate and ultimate causes of variation), a confusion that was reinforced by the first confusing use of the term by the French neo-Lamarckians. Indeed, with Darwin I have distinguished two epistemic uses of plasticity (as an explanans of variation *and* as an explanandum from the point of view of evolution) resulting from his clarifications about variation and his own uses of plastic- terms. But the French neo-Lamarckians introduced confusion between the ontogenetic or physiogenetic use of the term on the one hand and its phylogenetic or evolutionary use on the other hand. This was due to the central place they gave to plasticity within their conception of evolution while confusing its active and passive meanings. Therefore, the use of plastic- terms in evolutionary biology could have led to epistemic clarifications—and, in a way, its polysemy makes it a good candidate for a boundary concept—that allows the articulation of different types of explanations and universes of discourse (Darwin also used the term in its vernacular meaning given by breeders). However, the example of the French neo-Lamarckians also shows that if polysemy is not clarified, the boundary becomes blurred and confusion sets in, blocking the progress of knowledge. These persistent risk of confusion still accompany the different uses of the term within evolutionary thinking and biology in general.

I develop this idea of a boundary concept in chapter 8 and how plastic-terms might be used in this way in current evolutionary thinking, including developmental biology. However, I argue that we first need to be aware of the past confusions surrounding the use of plastic- terms in order to understand their current epistemic uses.

I now examine a first attempt to clarify the various uses of plastic- terms in biology within the context of the emerging evolutionary synthesis. The work analyzed here is not that of one of the fathers of the well-known modern synthesis, which gave the foundations for our current evolutionary thinking, but is that of an author whose work is a landmark both as a defender of Darwinism and as a precursor of a commitment toward a synthesis in evolutionary biology. His attention to the concept of plasticity is therefore of particular interest.

Plasticity, a Concept for a First Evolutionary Synthesis in Biology

In 1930 the British embryologist and evolutionist Gavin Rylands de Beer (1899–1972) published a small book, *Embryology and Evolution*. This influential book was the first to openly criticize the once popular theory of recapitulation (see chap. 2) of Ernst Haeckel. It is also the first work to highlight the rampant polysemy of the term *plasticity* in biology and the theoretical difficulties that can arise from it. Gavin de Beer explicitly discusses one of the main problems associated with the concept of plasticity: "A difficulty also arises from the fact that the term 'plasticity' has been used in many different senses" (De Beer 1940, 82). However, this difficulty did not prevent him from using the term extensively in his book. On the contrary, he attempts to clarify the various uses of the term through biology. The book reveals the particular understanding that scientists had of plasticity at a time when Darwin's theory of evolution was beginning to be accepted by biologists and when genetics was in its infancy. Before analyzing how de Beer defined plastic- terms in biology, it is important to set the theoretical context in which de Beer published his *Embryology and Evolution* in 1930. To do this, a brief overview of British embryology is warranted, particularly in the ways in which it differed from the well-known and influential German school.

It was not until the 1860s and 1870s that British embryology took off, well after Germany (where this development dates back to the early nineteenth century). One of the most prominent authors of early British embryology was Edwin Ray Lankester (1847–1929). A disciple of Darwin, he became the leader of transformist morphology in Britain. He carried out some descriptive embryological work based on the model and principles established by the German embryologist Ernst Haeckel a few years before him. Lankester's approach was to observe the embryonic development of many animal species, compare them, and deduce the evolution of corresponding lineages.

After Lankester, Edwin Stephen Goodrich (1868–1946), a trained embryologist, became the leader of evolutionary embryology in Britain in the 1920s. Goodrich emphasized the role of selection in animal morphology, studied the phenomena of mimicry,[4] and began to reflect on the difference between *modification* (change in response to external stimuli) and *mutation* (change due to a shift in the developmental or "germ constitution") (see Waisbren 1988). He is also one of those who came up with a rather sharp criticism of Haeckel's theory (see Ridley 1986, 62). Goodrich's student was Julian Huxley (1887–1975), who showed little interest, at least initially, in the phylogenetic dimension of embryology. But he was particularly impressed by the experiments of Charles Manning Child (1869–1954) on embryonic regeneration (the plastic ability of a developing organism to rebuild damaged parts of its body) and the work of Hans Spemann on embryonic induction (see chap. 2). Huxley soon became interested in the possible links between morphology and genetics and collaborated with the geneticist Henry Ford (1901–1988). Together they developed the idea of *rate genes*[5] from work on *Gammarus* (crustaceans) (Huxley and Ford 1925). They showed that by changing temperature, the gene that controls eye color works by altering the rate of certain chemical processes. At the time, this approach was relatively unique in the field (with the exception of research by the German geneticist Richard Goldschmidt— see Gilbert 1988, 311–346), since genetics was still mainly concerned with the study of gene transmission and the mapping of chromosomes (e.g., the work of the geneticist Thomas Morgan in the 1930s). The work of Ford and Huxley would lead to a broad reflection on the equilibrium between external and internal factors during ontogeny.

Consequently, in the late nineteenth and early twentieth centuries, British embryologists, unlike their German and American counterparts, tended not to radically oppose descriptive embryology, which was associated with phylogenetic study (marked by Haeckel's theory), and experimental embryology. The relationship between embryology and genetics differs markedly from that which prevailed in the great American "school" of genetics of Thomas Morgan and his disciples, which instead recommended separating the study of the transmission of genes (hereditary or phylogenetic studies) from that of their mode of action (studies of experimental embryology). British embryologists, on the other hand, tended from then on to promote synthetic approaches to the study of development, which brought together (and attempted to articulate) the work done in genetics and embryology, and a holistic or organic rather than a reductionist view of development. It is in this particular context that Gavin de Beer's original work must be seen.

If the "collective of thought"[6] that de Beer formed with Julian Huxley, and in particular their shared interest in experimental embryology, led him to publish a book with his professor entitled *The Elements of Experimental Embryology* (Huxley and De Beer 1934), he also turned, unlike his mentor, to comparative embryology. In this singular orientation one can see the influence of another of his professors: Goodrich (De Beer and Goodrich 1938). While Huxley had chosen to focus on many subjects other than morphology, de Beer devoted most of his research to the relationship between morphology and its implications in terms of adaptation, taking into account both hereditary and evolutionary connections (see Barrington 1973; Ridley 1981; Hall 2000). His interest in classical zoology led him to publish his 1930 book (*Embryology and Evolution*). The work contained in the improved edition was republished in 1940 under the new title *Embryos and Ancestors*. This book would be of considerable importance in the history of the relationship between developmental biology and evolutionary biology. In fact, it remained a reference in the field for almost fifty years, until the publication of *Ontogeny and Phylogeny* by Stephen Jay Gould in 1977 dealt the final blow to the Haeckelian theory of recapitulation, which held that the development of the animal embryo followed the evolutionary development of the species.

De Beer begins *Embryos and Ancestors* by recalling the history of the notion of the *scala naturae*, or "scale of beings," and its homology with the development of the organism. From the first chapter, de Beer openly attacks Haeckel's "theory of recapitulation" (also known as the "law of fundamental biogenetics") and describes a general opposition, which he believes has been observed throughout the history of the life sciences, between those who might have considered ontogeny to be the cause of phylogeny (notably Walter Garstang [1868–1949]) and those who might have consider phylogeny to be the cause of ontogeny (the proponents of Haeckel's theory). At the end of the first chapter of *Embryos and Ancestors*, de Beer announced the aim of his work: to make a synthesis of the current knowledge of development, evolution, and heredity in order to find a harmonious formula capable of coordinating all these data (De Beer 1940, 9). De Beer therefore proposes to take the problem back to its starting point and to develop a synthetic view of the whole field of biology. According to him, the answer to the question of the relationship between phylogeny and ontogeny requires reference to the mechanisms of heredity. To reach this conclusion, he relied on the work of Ernst Mehnert (1864–1902), Louis Agassiz, and Franz Keibel, who pointed out that some characters of the organism show a specific order of appearance in phylogeny, an order that is not always respected in ontogeny (Mehnert 1898). For instance, de Beer says, "Teeth were evolved before tongues, but in mammals now tongues develop before teeth" (De Beer 1940, 6).

To get around this problem of the order of appearance of characters, de Beer points out that changes should be observed not in the stages of development or in the stages at which characters appear on the scale of beings but directly in the characters themselves. The reference to characters rather than stages makes the link between genetics, embryology, and evolution possible, since characters, unlike stages, are not measured differently depending on the field of study. In this way, de Beer effectively uses characters as a methodological boundary concept that allows multiple fields to refer to and inform about the same phenomenon. The emphasis on characters is also why de Beer rejects Haeckel's recapitulation theory, which is based on stages defined in embryology. De Beer shows that each stage can only be seen as an arbitrarily selected "section" of the time axis of an organism's life. It cannot therefore be seen as a "reference point" or "gold standard" valid for all fields of study within biology. A "stage" in embryology has a meaning for that field and cannot necessarily be translated into a similar stage in evolutionary biology or even genetics (see Nicoglou 2017).

Furthermore, by focusing on characters rather than stages, de Beer creates a new methodological paradigm—in the Kuhnian sense of an epistemic revolution—and introduces a new methodology into biology. The new methodological approach will allow some characters of the organism to be isolated from their temporal context, from the period or stage in which they have usually been observed or described, especially in embryology. By focusing on the characters, de Beer sets a new standard—common to genetics, embryology, and phylogeny—that allows the scientist to disregard the temporal context. De Beer was the first to integrate the study of genetics with that of phylogeny by focusing on characters, which were then defined in the same way in all fields of biology (Hall 2000, 726).

It is in this highly innovative context that de Beer takes up the concept of "plasticity" and highlights the difficulties posed by the polysemy of the term. First of all, he sets out to list the different meanings that the term assumes according to his understanding of its use in different fields of study. In this way, the definitions he proposes are original, as de Beer does not refer to the literature within the disciplines for his definitions. He simply seeks to highlight some of the nuances that exist between the different meanings of *plasticity*.

In the book he begins with what he calls "genetic plasticity":

> Genetic plasticity is the ability of individuals of a species to show a high degree of variance, and this condition obtains when the number of individuals carrying genes in the heterozygous condition, and the number of those genes, is large. In such species the possibilities of recombination of genes are numerous, and there is a reserve of recessive genes which may come into play, in one way or another, in the new conditions which recombinations and permutations provide. (De Beer 1940, 82)

Thus, "genetic plasticity," a result of the maintenance of genetic diversity within the species, allows individuals of the species to vary greatly from one another. De Beer thus links "genetic plasticity" to the issue of variation, which allows him to establish a direct link between variations in character (phenotypic variations) within a species and the genetic modifications associated with them.

De Beer distinguishes the genetic sense of plasticity from another, more physiological, sense, which he calls "histogenetic plasticity," referring "to the ability of a tissue to undergo further or other differentiation" (De Beer 1940, 83). Whereas genetic plasticity refers to the degree of character diversity based on the conservation of genetic options, "histogenetic plasticity" refers to the degree to which tissues can differentiate or specialize. Regeneration experiments in tissue culture have shown a direct correlation between the developmental stage of the tissue and its histogenetic plasticity. The earlier cells are taken in development, the faster they can grow or switch to a new differentiation pathway. Conversely, if the cell is taken at a late stage of embryonic development, it is irreversibly locked into a differentiation pathway that does not allows it to change its future fate. "In other words, the histogenetic plasticity of the tissues which characterizes the young stages of development is lost at the older stages" (De Beer 1940, 83). This "histogenetic" type of plasticity is the one most closely associated with the use of plastic- terms in embryology at the same time, particularly with Spemann (see chap. 2), a meaning that is relatively similar to its use in stem cell biology today (see chap. 6).

Finally, de Beer proposes a third definition of plasticity (the first one presented in the book), which he calls the "phylogenetic or evolutionary plasticity," and which he somewhat enigmatically describes as "the potentiality [for species] of evolving further" (De Beer 1940, 81). De Beer observes that the higher levels of the animal kingdom (i.e., phyla or classes) contain more diverse types within them than do the smaller groups (i.e., families, genera, or species). His hypothesis is that the changes that gave rise to the large groups allowed the animals in the large groups to evolve further and in more directions (i.e., to be more evolutionarily plastic) than the animals in the small groups. De Beer thought that major evolutionary changes, and therefore the large groups of the animal kingdom, were due to paedomorphosis (the appearance of characters in early stages of ancestral ontogeny, when there is still high histogenetic plasticity). Consequently, de Beer argues,

> Since phylogeny is the result of changes in successive ontogenies, it is impossible to expect much alteration to take place when the animal has reached the later stages of its development [because histogenetic plasticity decreases over time]. This is presumably why gerontomorphosis [changes in the adult organism as a result of aging] can only result in the production of small groups of animals, which become more and more specialized and incapable of evolving further. (De Beer 1940, 83)

Having considered the relationships between histogenetic and phylogenetic plasticity, de Beer sets out to understand the relationships between histogenetic, phylogenetic, and genetic plasticity. An important question is whether histogenetic plasticity, when prolonged—for example in cases of neotony—can play a role in correlatively increasing genetic and evolutionary plasticity. The view he proposes opens up the possibility of an original rapprochement between genetics and embryology. Such a proposal, which would consist in comparing histogenetic plasticity and genetic plasticity in order to study, in a transversal or interdisciplinary way, the different data on character plasticity, will be precisely the main agenda of what was soon called "developmental biology," whose purpose is precisely to identify the (genetic) determinants of differentiation and morphogenesis. In short, and to use de Beer's categories, developmental biology studies and analyzes the relationships that exist between "genetic plasticity" (the genetic diversity within organisms) and "histogenetic plasticity" (their expression through cell and tissues polymorphism). Developmental biology emerged in the late 1930s from the work of Gluecksohn Waelsch (1907–2007) and Conrad Hal Waddington (1905–1975), who sought to clarify the origin of mutations affecting early stages of development. Both were able to identify the role of the genes involved by observing altered developmental processes. At this time, developmental biology became the branch of biology that studies the processes by which organisms grow and build themselves and in particular the genetic control of cell growth, cell differentiation, and morphogenesis.

At this point, de Beer concludes his analysis of plasticity concepts with the idea that different disciplines of biology are linked by the concept of plasticity (see the epigraph to this chapter).

De Beer was somehow convinced that the different uses of the concept of plasticity in different fields of study made it possible to bring together the findings from these different fields in order to offer a synthetic view of biological phenomena. The above conclusion takes the form of an advice, or even an exhortation, from de Beer to his successors. Unfortunately, it remained without effect, because in the synthetic view of evolution that has been elaborated since the 1940s, embryology lost favor thanks to the loss of focus on developmental mechanisms.

De Beer's constant effort was to synthesize knowledge from different fields of biology (i.e., embryology, genetics, and evolution) into a unified view. To this end, he defined and described a family of concepts of plasticity from the perspective of these three subfields. However, it was never his intention to "dissolve" these fields into a single, large, overarching field whose aims and methods would be unified or to elevate one field above the others. Thus, for de Beer, if comparative anatomy should not impose its criteria for understanding all mechanisms, as Haeckel had tried to do for many years by criticizing with some irony the work of experimental embryologists, neither could genetics alone replace embryology for explaining morphology, just as experimental embryology is not sufficient to explain the homologies that exist between different organs. For these reasons, de Beer really did use the concept of plasticity as a boundary concept, *a concept at the boundaries* of biological fields to somehow favor transdisciplinary work.

De Beer can be seen as the first initiator of an evolutionary synthesis, bringing together data from genetics, embryology, and evolution. However, it was not until the 1940s that a formal theoretical synthesis emerged in evolutionary biology. This synthetic conception soon became the main paradigm in biology for understanding evolution. I will now examine the role of plasticity within this theoretical synthesis in comparison with the role it had in de Beer's synthesis.

The Modern Synthesis, or the Hegemony of Genotypic Plasticity

In genetics, the debate between biometricians and Mendelians, which began in the 1910s, continued for some twenty years. On the one hand, the Mendelians argued that there was a close relationship between the mode of genetic

PLASTICITY IN EVOLUTIONARY BIOLOGY

variation (discontinuity) and the mode of evolution (saltationism), with
evolution being discontinuous and achieved by infrequent but rapid spread
of variation within a population. On the other hand, Darwinian naturalists
were convinced that evolution was gradual and continuous, a slow accumulation of traits. Furthermore, whereas Mendelian naturalists were interested
in the mathematical behavior of the elementary units of heredity—genes as
units transmitted from parents to offspring—Darwinian naturalists were
concerned mainly with population phenomena and envisaged holistic interpretations (not in terms of probabilistic measures but in terms of statistical phenomena).[7] With the development of population genetics between the
1920s and 1940s, the debate was thought to have been "settled" (the idea that
this debate was settled with the modern synthesis is still highly controversial;
see Pigliucci and Müller 2010).

At the beginning of the twentieth century, the zoologist Maynard M. Metcalf (see chap. 3) believed that it was possible to report "well-defined trends
in evolution." He had carried out crossbreeding experiments to see whether
variations and mutations all tended in a particular direction or, conversely,
followed a balanced distribution around an average. These were the beginnings of the first population genetic studies. Fisher, Haldane, and Wright
made the most important contributions to the field, relying in particular on
mathematical models. In promoting population genetics to the detriment of
embryology or what was to become "developmental biology," these population geneticists—also advocates of a synthetic view of evolution—left aside the
notion of "plasticity" and its various uses. Their "synthetic view" mostly subsumed other subfields under the banner of population genetics. At most, they
retained the reference to de Beer's notion of "genetic plasticity" by focusing
on issues related to the "norm of reaction" (chap. 3), which Dobzhansky later
popularized by referring to the precise notion of plasticity (Dobzhansky 1955).

Similarly, these theorists paid little heed to de Beer's recommendations
that different areas of biology should be given equal importance in order to
achieve a fully synthetic framework. They focused their attention on how,
first, Mendelian genetics could shed light on natural selection, and second,
on how population genetics could explain natural selection without giving
equal attention to how these fields of study might be articulated with embryology. In a way they seem to have left aside the question of variability,
which had mostly been studied with embryology, in favor of the study of
heredity, which was at the center of the emerging field of genetics. Perhaps
they had misinterpreted Darwin's words when he claimed that "any variation
which is not inherited is unimportant for us" (Darwin 1861, 13). In fact, Dar-
win thought that variation due to variability was central to the evolutionary

process but that variation had to be inherited to be part of natural selection mechanisms. Both principles were equally central to Darwin. On the other hand, for the theorists of synthesis, inheritance seems to be more important than variability in the phenomenon of natural selection—which was convenient at the time, because the laws of inheritance had been discovered since Darwin while those of variability remained to be understood. Embryology therefore remains outside the core of the modern synthesis.

The issue of the articulation of different fields of study within the theoretical synthesis of evolution is an essential point for understanding the structuring of biology during this period and how evolutionary questions were formulated afterward. For the theorists of the "synthesis," the problem was no longer to answer general questions about the origin of characters or variations (as it had been for the naturalists). They wanted to understand the specific mechanisms at work in nature as a result of natural selection. To this end, in parallel with the emerging stream of "theoretical synthesis" in evolution, several fields of study, themselves subdivided into subfields, were used (e.g., genetics was subdivided into Mendelian genetics and population genetics and later into other subfields).

As evidenced by contemporary definitions of what Huxley called the "Modern Synthesis" in 1942, the synthesis is then thought to be the integration of the Mendelian heredity and population genetics with Darwinian theory. Fisher, Haldane, Wright, Dobzhansky, Huxley, Mayr, Rensch[8] (1900–1990), Simpson[9] (1902–1984), Stebbins[10] (1906–2000) and others carried out this synthesis in the 1930s and 1940s. Of these, only Dobzhansky and Huxley were interested in embryology at the beginning of their careers, although both abandoned it for the study of genetics and its relationship to evolution. Their knowledge of embryology would therefore have little influence on the initiation of an evolutionary theoretical synthesis. Consequently, such a "synthesis" was only partial compared to what de Beer had proposed. At the very least, it did not integrate the three fields of genetics, embryology, and evolution on an equal footing.

This "segmentation" of fields and subfields of study around more precise questions and the neglect of variability understandably led to the rejection of such a vague term as *plasticity*. Indeed, whatever precise meaning one might ascribe to it, plasticity remained a very vague term covering the broad idea of variability in living bodies (whatever they might be), the specific causes of which remained to be explained.

But my argument goes further. The disappearance of the "variability" problem is not due to one sentence by Darwin in the On the Origin of Species. I argue that a possible link between Darwin's view of variation and Lamarck's idea of the inheritance of acquired characters also led to another direction, namely genetic inheritance. In both Darwin's and Lamarck's theories there is the idea that individual differences are the true heritable elements or "characters," strictly speaking. This is why Darwin's theory can be described as a theory of "generation" in the old tradition, which does not make a clear distinction between heredity and embryonic development. Wilhelm Johannsen had called this conception the "transmission-conception" of heredity or the conception of "apparent" heredity (Johannsen 1911). This relationship between development and inheritance is clear because Darwin does not distinguish between ordinary variations and monstrosities. For him, all variations, large or small, are individual deviations from the "pattern" of normal development. The question of the causes of variability (and thus of plasticity) was of paramount importance to Darwin but not to his successors, for whom variation is virtually "unlimited" because random mutations at the level of populations are seen as the underlying cause of change.

Alfred Russell Wallace (1823–1913) first introduced a more population-centered view of variation (although in a sense it is difficult to say that Darwin's conception was not also a population view of variation), but its extension within a theoretical synthesis of evolution is mostly the result of the interaction of evolutionary biology with genetics. When Johannsen introduced the concepts of genotype and phenotype in 1911, he not only provided a new terminology for the nascent field of genetics but also suggested that Darwin's and Lamarck's "individual differences" were actually due to the "reactions of the gametes" when they came together to form a zygote (Johannsen 1911, 130, 140). In other words, he showed that the nature of variability (the difference in expression of the phenotype) was not determined by the transmission of individual differences from parents or ancestors to offspring. Johannsen calls his conception the "genotypic conception" of heredity (as opposed to the transmission conception), which somehow managed to erase variability in favor of heredity. He demonstrated this new conception of heredity by carrying out crosses on pure lines, following Mendel's example. Genetic mutation became central to this view. Thus, for William Bateson, Hugo de Vries, and Thomas Morgan (all three proponents of mutation theory[11]), there was a clear distinction between heredity and individual development (Bowler 1989).

From the "rebirth" of Mendelian genetics in the early twentieth century and its theorization, "it was no longer possible to believe that a character [or a variation] acquired during ontogeny (whether adaptive or merely some 'accident of growth') could be inherited so as to form the basis for evolutionary change" (Bowler 2005, 19–20), because individual development does not have the ability to adjust the heritable aspects of the genome. Germ line rearrangements seem to occur indirectly, producing a range of new and different characters. Mutation geneticists (who believe that mutation is based on new combinations rather than replication errors) thought that these germ line rearrangements were the basis of the natural selection process. In a sense, this new explanation would complement the theoretical problem of the causes of variation that remained when Darwin introduced the concept of plasticity.

By linking this new explanation of germ rearrangement as the basis of evolution with the old Weismannian theory of germ plasma continuity as the basis of heredity,[12] the problem of variation in natural selection shifted definitively to questions of heredity rather than to questions of development (and hence plasticity). As Bowler noted, the Weismannian view would change "the logic of the selection theory by suggesting that determinants for a wide range of characters persist [not to say, preexist] within any normal population" (Bowler 2005, 20). In this view, variation appeared as a mirror phenomenon of inheritance. All variability was predetermined by the rigid inheritance of determinants within the population. The new characters appeared through germinal transformation (rather than through plasticity, as was the case for neo-Lamarckians) and were then preserved by heredity (Bowler 2005).

The biometric school, the mathematical study of heredity (the descendants of Mendelian thinking), had shown the existence of a range of natural variation in wild populations and the occurrence of microevolution through the action of natural selection on this variation. Fisher's mathematical models in the 1930s led to the classical genetic theory of natural selection when he showed that "variation in a wild population is maintained by the circulation of genes through sexual recombination." Fisher also showed "how [this] variability could be shaped by natural selection [if] some genes reproduced more efficiently than others because they conferred a slight adaptive benefit" (Bowler 2005, 23). In other words, with this new view, it was definitely no longer necessary to interpret variation at the individual level in order to explain natural selection at the population level, as had been the case for Darwin.

Because of this new conception of variation, the theoretical synthesis of evolution no longer needed plasticity, because in this view, and in contrast to Darwin's view, variability was no longer a key issue in explaining variation. However, this concept of variation is profoundly incomplete because it does not explain all of the variability in nature or take into account the laws of variability, including those that rely on mechanisms other than germ recombination to induce variation that may be heritable.

Weismann and Galton (the biometricians), in turn, by restricting the range of variation to germinal (hereditary) and somatic (nonhereditary) variation, would ultimately undermine the plausibility of variation and evolution as the results of changes in the developmental pathway, which for Darwin (Darwin [1868] 1872) were direct consequences of "organisational plasticity." Variability (or plasticity), in their view and that of the theorists of the synthesis, depends solely on the hereditary transmission of genetic recombination, which is studied by genetics.

Despite ongoing controversy between neo-Darwinians and mutationists (including their successors, the Mendelians), the suppression of the developmental model of heredity and variation facilitated the conceptual framework within which a synthesis could be developed. In this synthesis, variation consists of alternative genetic factors maintained in the population by recessive, heterozygous[13] inheritance and potentially subject to selection. The combination of population genetics with the study of geographical variation will bring together two lines of interest that are central to the understanding of natural selection processes. From then on, variation was no longer seen as a force for change distinct from heredity, often referred to in the past as plasticity. Variation is now represented by genetic diversity, maintained in the population by heredity and subject to selection. Its measurement no longer requires an understanding of phenogenesis (at the individual level) but an understanding and interpretation of genetic distributions (at the population level).

This theoretical synthesis in evolutionary biology is further developed and refined under the name of the modern synthesis of evolution. The work of William Donald Hamilton (1936–2000), George Christopher Williams (1926–2010), John Maynard Smith (1920–2004) and others led to the development of a genocentric view of evolution in the 1960s. Since its beginnings in the 1930s and 1940s, the modern theoretical synthesis of evolutionary biology has thus continued to expand to include new scientific discoveries and new concepts unknown at the time of Darwin. This expansion has led to increasingly rigorous analyses, sometimes formalized in mathematical models, of phenomena as diverse as kin selection, altruism and, of course, speciation. Conceptually, variation will be more closely associated with questions of heredity than with questions of development. Moreover, the identification of hereditary factors with genes implies that the processes that cause variation are the results of sexual recombination or errors in gene replication, the mechanisms of which are peripheral to the question of evolution. In this new theoretical scheme, the notion of plasticity seems completely obsolete.

It is also claimed that there is a direct link between the adaptation of a phenotypic trait and the presence of genes associated with the expression of that trait. Therefore, any adapted variation is associated with a difference in gene expression. More often—and although such a view would soon be criticized because of the need to simplify the relationship between gene expression and phenotypic character—the link becomes a kind of tacit assimilation between genetic factors and the phenotypic result. Consequently, the notion of phenotypic plasticity—the ability of an organism to express different phenotypes from a given genotype under different environmental conditions—will soon replace de Beer's notion of genetic plasticity—the ability of individuals of a species to show a high degree of variation. The notion of phenotypic plasticity therefore presupposes a genocentric view.

Another central factor in Darwin's work that the modern synthesis ignores is the role of the environment in evolution. In the modern evolutionary synthesis, the environment is abstracted into an "optimal environment," and its effects are generally neglected in comparison to genetic information. In fact, the wordy definition of evolution as changes in the frequencies of the collective genetic resources of populations of those genetic programs that lead to the successful development and survival of individuals under favorable environmental conditions (Dobzhansky and Boesiger 1983) is less influential and widespread than the more direct and concise statement that "evolution is a change in genetic frequency" (Dobzhansky [1937] 1982), which omits the environment altogether.

On this view, biologists would tend to see the environment as relatively constant and focus primarily on the role of mutations in the evolutionary process. While the environment can be seen as a source of perturbation for evolutionary analysis, mutations are *the only real causes* of variation. Statistics and its mathematical formalisms have a privileged position in these studies and contribute to the neglect of the complex interactions between genotype and environment.

When Fisher introduced the analysis of variance (ANOVA) in 1918, it was a convenient way for him to decompose phenotypic variability (or plasticity) in a given population into its genotypic and environmental components and their interactions. Wright (1920) later offered some more technically refined, but in principle identical, methods. In the statistical view they developed after 1950, heritability in the broad sense was seen as the proportion of phenotypic variance that was due to genotypic variation. The rest of the variance was attributed to environmental variation, which was assumed to have no direct effect on heritability. It is important to note that this view relies on the assumption that the interaction between genotype and environment is the same for all genotypes and all environments during development and that analyses based on a single moment in the lives of the individuals in the sample are representative of the population. In such a view, the validity of the model is ensured by the normative idea that there is a constant regularity of gene-environment interaction.

One consequence of this analysis is that it ignores the plasticity of the genotype-environment interaction, assuming that the genotype-environment interaction always remains unchanged. This would have consequences: by not taking this kind of plasticity into account, biologists began to understand the idea of genotype-environment interaction often too rigidly. It was precisely to counteract this overly rigid conception of genotype-environment interaction that the notion of "norm of reaction" emerged as a powerful conceptual tool in the Soviet Union in the 1920s (Sarkar 1999). As it spread through the scientific literature, the concept became an indicator for each individual phenotypic curve and was conceived as the result of the "reactions of the genotypic components" in contact with different environments (see chap. 3). The gradual modification of the concept after its introduction in the English-speaking world by Dobzhansky in 1937 (see chap. 3) and its assimilation into the notion of phenotypic plasticity (see chap. 5) led the twentieth-century biologists to interpret plasticity only in the sense of a genotypic plasticity (soon thereafter called, ironically, despite its genocentric assumptions, phenotypic plasticity). Henceforth, the term *plasticity* refers only to the idea of "the property of a given genotype to produce different phenotypes in response to distinct environmental conditions" (Pigliucci 2001, 1).

From the beginning of the modern synthesis, the issue of the complexity of the genotype-environment interaction over the life course of the individual is no longer taken into account. This idea persists to the present day as evidenced by Pigliucci's definition, which focuses mainly on the fact that there is an interaction between genotype and environment and that the phenotype is the result of this interaction but does not take into account the plasticity of this interaction itself.

So what happens to plasticity in this new context? De Beer had proposed two different categories of plasticity corresponding to these two fields: histogenetic plasticity (which includes cells, tissues, organs, and whole organisms) and genetic plasticity. Based on my previous analysis, it seems that the modern synthesis considers only genotypic plasticity, and the prevalence of the term can be observed from the 1980s onward. Population geneticists investigated other categories of plasticity only insofar as it was thought that all forms of plasticity could be analyzed genetically. Consequently, the only problem that a biologist of the modern synthesis paradigm could address in relation to plasticity was to know how plasticity evolved, since they thought that plasticity, like any other trait, must itself be subject to natural selection. We will examine this in the next chapter.

5

Defining *Plasticity*

Plasticity is therefore shown by a genotype when its expression is able to be altered by environmental influences. The change that occurs can be termed the response. Since all changes in the characters of an organism which are not genetic are environmental, plasticity is applicable to all intragenotypic variability.

ANTHONY BRADSHAW (1965, 116)

This chapter returns to the precise roots of Anthony Bradshaw's formulation of the notion of "phenotypic plasticity" in genecology and its gradual adoption by biologists more generally. It describes how the concept of plasticity came into operational use in the 1960s and how, as a result of the modern synthesis's conflation of the terms, evolutionary biologists from the 1960s onward began to discuss plasticity from the perspective of genetics.

Phenotypic Plasticity

In 1965 Anthony D. Bradshaw (1926–2008) published an article titled "The Evolutionary Significance of Phenotypic Plasticity in Plants." In this article he proposed a new model, based mainly on his studies of plant populations in California, to explain the evolution of norms of reaction. Bradshaw points to the importance of environmental influences on organisms: "We are becoming increasingly aware that the individual cannot be considered out of the context of its environment" (Bradshaw 1965, 115). Before him, the "instability" produced by the environment had mostly been seen as a source of perturbation for geneticists (Falconer 1952). In 1965 Bradshaw claimed that "any modifications induced by the environment during the course of an experiment are usually considered only an embarrassment" (Bradshaw 1965, 148). Therefore, geneticists have usually tried to eliminate it in their studies. However, the problem remains of understanding how the individual can maintain some stability in an environment that is by definition "unstable."

After the Second World War, the question became one of the main topics for geneticists (e.g., Dobzhansky and Wallace 1953; Mather 1953; Jinks and Mather 1955; Kimura 1955; Lerner 1954; Lewontin 1957; Waddington 1959;

Levins 1963). Bradshaw, for his part, incorporated the conclusions of the Nilsson-Ehle study (1914) to illustrate the fact that "instability"—or rather "plasticity," as he chose to call it after Nilsson-Ehle[1]—could also have an adaptive value in many circumstances. But although Bradshaw followed Nilsson-Ehle's conclusions about the adaptive value of plasticity,[2] his interpretation of the genetically determined nature of plasticity was quite new. Bradshaw argued that plasticity should be studied as such and defined it as follows: "*Plasticity* is therefore shown by a genotype when its expression is able to be altered by environmental influences. The change that occurs can be termed the *response*. Since all changes in the characters of an organism which are not genetic are environmental, plasticity is applicable to all intragenotypic variability" (Bradshaw 1965, 116).

In Bradshaw's view, plasticity is highlighted by the potential alteration in the expression of the genotype under the effects of environmental influences. The "response" to environmental change can itself be understood as an adaptation. Plasticity therefore embodies the potential phenotypic variability associated with the genotypic response to an environmental change. Indeed, Bradshaw believes that a genetic basis determines plasticity (Bradshaw 1965, 145). Thus, unlike Nilsson-Ehle, Bradshaw viewed "adaptive plasticity" not only from an evolutionary perspective but also from a neo-Darwinian perspective.

If the formulation of the notion of norm of reaction in the early twentieth century can be understood as a key notion in the history of Mendelian genetics, a notion that involves a mitigation of genetic determinism, then with Bradshaw the notion of plasticity becomes part of the history of quantitative genetics. The overview he offered of plasticity in a genetic-ecological perspective opened up the possibility of a precise analysis of its evolution.

However, if Bradshaw's 1965 review of plasticity opened up a new field of study—the study of the evolution of plasticity with optimality models—for further research in the 1980s, in the 1965 article he mostly discussed his predecessor's positions. It seems, however, that some of these discussions may have been confusing to quantitative geneticists, whose interest in plasticity only began in the 1980s (I analyze these biologists' understanding of the evolution of plasticity in the next section of this chapter).

In the 1965 article Bradshaw distinguishes between what he calls "physiological plasticity"—which represents all forms of plasticity[3]—and "morphological plasticity"—which Bradshaw saw as a particular manifestation of physiological plasticity that associated with the changes that occur during development. Such a distinction is somewhat reminiscent of Goldschmidt's distinction between the "norm of reaction" and the "norm of reactivity"

(Goldschmidt 1940, 250). However, whereas Goldschmidt used the distinction to emphasize the genetically determined nature of the "norm or reaction" as opposed to the "norm of reactivity" (which would apply to any physiological reactivity of the organism), Bradshaw equated "physiological plasticity" with "morphological plasticity" and considered both to be genetically determined. According to Bradshaw, both were "adaptive plasticity." So the only reason for continuing to distinguish between the two notions—morphological plasticity versus physiological plasticity—was to make it clear that the type of plasticity that biologists were mainly concerned with was "morphological plasticity" because that was the type of plasticity that was most easily identified in nature. Take, for example, the desert locust *Schistocerca gregaria*, which expresses two different morphological phenotypes depending on the environmental conditions in which it develops (Marriott and Holloway 1998). It will be green with small wings if food resources are limited during its development. It will be dark with long wings if food resources are more abundant. It should be noted that contemporary biologists largely reject Bradshaw's distinction between physiological plasticity and morphological plasticity since they no longer use this kind of terminology and have focused instead on phenotypic (or, more recently, developmental) plasticity.

Indeed, after Bradshaw's work, few studies in quantitative genetics have focused on physiological plasticity—and those that do employ a different meaning of the term than Bradshaw's.[4] As a result, Bradshaw's distinction (between physiological and morphological plasticity) may now seem a source of confusion to biologists who are reading and referring to his work.

To clarify his definition, Bradshaw further distinguished between plasticity and a phenomenon called phenotypic flexibility, which he argued should not be confused with plasticity. Thoday (1953) referred to phenotypic flexibility in 1953 to characterize the ability of an organism to continue to operate within a given range of environments. In Bradshaw's view the difference between plasticity and flexibility rests on the following nuance: whereas plasticity refers to the *variability* of the norm of reaction, flexibility may also include stable responses to environmental change, where the reaction norm is not an issue because there is no phenotypic change. For Bradshaw, flexibility had nothing to do with norm of reaction because there was no actual phenotypic change. Therefore, for Bradshaw the phenomenon of heat resistance is not a case of plasticity but only a case of phenotypic flexibility. However, Bradshaw's position itself was not clear in the article, as he also claimed that phenotypic flexibility could occasionally involve both stable and plastic responses (Bradshaw 1965, 117). This makes it difficult to assess the difference between what is adaptive and what is not. This problem became the subject

of some controversy over the definition of *phenotypic plasticity*, especially as
it related to its evolution.

Bradshaw's view that plasticity is genetically determined puts him at odds
with those who believe that the plasticity of a trait should be linked to the specific development of the trait. Bradshaw based his convictions on the common idea among (plant) embryologists that traits that require meristematic[5]
development (such as leaves) would be more plastic than traits that form
quickly and directly (such as sexual organs). In his view, this idea is flawed for
the following reason: if you compare the relative plasticity of certain traits in
different species but with similar modes of development, you should find the
same relative distribution of plasticity in these different species. This is not
the case. He argues that such results would only be expected if plasticity was
actually linked to the development of the plastic trait. For similar reasons, he
also rejects the idea that plasticity is a property of the entire genome. His view
on plasticity is based on the following three points:

(1) Plasticity is a property of individual characters in relation to specific en-
    vironmental influences.

(2) Since the plasticity of a trait varies among different species of the same
    genus and among the different varieties of the same species, the genetic
    determination of plasticity and the genetic determination of the plastic
    trait are independent. He emphasizes that "it is difficult to explain such
    differences unless you accept the fact that the plasticity of a trait is an
    independent property of this trait and is subject to a specific genetic con-
    trol." (Bradshaw 1965, 119) (This point will be the source of further con-
    troversy concerning the evolution of plasticity.)

(3) Finally, he refers to the work of Waddington (1953) on canalization and
    genetic assimilation of characters and observes that if stability is geneti-
    cally determined, the same might be true of plasticity, which by definition
    is opposed to stability.

These three arguments allowed Bradshaw to defend the hypothesis that plasticity, being genetically determined, is necessarily subject to natural selection
and can evolve like any other trait.

In 2006, two years before his death, Bradshaw was invited to a symposium organized by the journal the *New Phytologist*, to speak on the subject
of plasticity forty years after his famous 1965 article. In what was to be the
last published article of his life, Bradshaw offered a fascinating perspective
on the origins of his ideas in the field (Bradshaw 2006). Among other things,
the article reveals that his position had not changed much since the 1965 article, or at least that it had been enriched by recent work in the field. Such
a view seems to reflect the "fixation" and "consolidation" of the concept of

"phenotypic plasticity" after Bradshaw offered it in his first review of the literature in the mid-1960s.

Thus, unlike the concept of "norm of reaction," whose definition was debated for a long time, the concept of "phenotypic plasticity" became, from its first appearance (in 1965, with Bradshaw),[6] a concept that we could say was *perfectly defined from a scientific point of view*. Several points are important to note here. First, from that moment on, most of the studies dealing with phenotypic plasticity explicitly refer to Bradshaw's seminal paper of 1965, which somehow signaled the birth of the concept in genetics. Second, and more importantly, as with Spemann's concept of induction or Johannsen's distinction between genotype and phenotype, Bradshaw based his definition of phenotypic plasticity on experiments he had carried out during his PhD (1952, 1954, 1959, 1964) and during his "sabbatical" in California (1964), where he joined Robert Allard (1919–2003) (Allard and Bradshaw 1964). During his sabbatical he also collaborated with Subodh Jain, with whom he published in 1966 the first of a long series of articles on "evolutionary divergence among populations of adjacent plants" (Jain and Bradshaw 1966). These reflections and the work done during the California trip led Bradshaw to focus on issues related to the interactions between the genotype and the environment (Fitter 2010). Third, Bradshaw offered a definition of phenotypic plasticity that remained stable (which is why it is no longer hotly debated): "Plasticity is therefore shown by a genotype when its expression is able to be altered by environmental influences" (Bradshaw 1965, 116).

The use of the concept of "phenotypic plasticity" increased in the scientific literature after the publication of Bradshaw's review article on plasticity in 1965. "Phenotypic plasticity" somehow became an operative concept in biology, and especially in genetics, where in the past, and especially in embryology, the term *plastic* had mostly a highly metaphorical and architectural connotation (see chaps. 1 and 2). This is largely due to Bradshaw's first attempt at an operational definition. However, the emergence of the concept in genetics is closely linked to the history of the field itself. While the role of the environment and its influence on inheritance remained important for Darwin and his theory of evolution by natural selection, the first geneticists of the early twentieth century quickly tried to distinguish the role of external factors from genetic ones. This attempt led to the opposition of two camps: the "stability" advocates and the "interactionists"—an opposition that lasted at least until the middle of the twentieth century. The first camp did not overlook the external factors but mostly tried to dissociate them from the genetic ones and put the emphasis on the latter (e.g., Johannsen, Bateson, Dobzhansky). The second looked at the interaction between genetic and external factors and tried to understand what in this interaction could be transmitted from one generation to the next (e.g., Woltereck, Hogben, Schmalhausen, Goldschmidt). But both camps were really concerned with the interaction of factors or with the apparent "plasticity" in the inheritance of traits. At some point, therefore, it remained important for geneticists to define clearly what was at stake in order to integrate it into either the stability or the interactionist view.

Bradshaw never saw the environment as a problem, as some of his predecessors did. One of the reasons for this is that Bradshaw was an ecologist before he became a geneticist. So he explicitly included the environment in his genetic analysis. However, Bradshaw's geneto-ecologist perspective did not close the debate on other interpretations of the evolution of phenotypic plasticity for very long. Indeed, it led to one of the most important controversies of the 1990s between the biologists Sara Via and Samuel Scheiner, which I will now examine. Via saw phenotypic plasticity as a by-product of evolution by natural selection, while Scheiner saw it as a trait like any other, something that is genetically determined and as such can be studied using the tools of population genetics.

With the advent of genetics and the work of Bradshaw, the term *plasticity* began to be used more frequently as an operative notion in genetics. However, the texts written by biologists trained in both embryology and genetics attempted to impose a multidisciplinary approach on the field at the time. This should convince us that the use of the term as it was used in embryology (chap. 2) would not disappear completely with the rise of genetics, since it meets specific requirements that are not necessarily met by the "genetic use" of the term (see chap. 7).

Since Plasticity Is the Result of Genetic Diversity, Plasticity Must Evolve

TESTING ADAPTIVE PLASTICITY

The question of the evolution of plasticity has been addressed mainly by population geneticists as part of modern synthesis trying to understand the role of natural selection on plasticity. Their research was in line with the work initiated by Anthony Bradshaw in the mid-1960s. Like him, they thought that physiological and morphological plasticity are on the same level in the sense that they both depend on the same kind of genetic control. Their focus is therefore mainly on what can be defined as *adaptive plasticity* (see chap. 7).

In 1985 Russell Lande and Sara Via produced the first quantitative genetic models of the evolution of *phenotypic plasticity* (the term now used to describe phenotypic response to the environment) (Via and Lande 1985). Quantitative genetics studies the genetic component that explains variation in quantifiable traits (e.g., size, coat color, growth rate, concentration of a molecule, etc.) and their heritability. With the advent of the synthetic theory of evolution, quantitative genetics became a common tool in biology (Fisher [1930] 1999; Wright 1949; Falconer [1960] 1981). Subsequently, quantitative genetics has attracted increasing attention in many related fields of evolutionary biology (Lande 1980; Cheverud, Rutledge, and Atchley 1983; Lande and Arnold 1983; Slatkin 1987; Barton and Turelli 1989; Shaw et al. 1995; Roff [1997] 2012).

Naturally, these techniques were used to assess the evolution of phenotypic plasticity (Falconer 1952; Via 1984a, 1984b;, Via and Lande 1985) in the context of work that opened up a new avenue of research into plasticity. The biologist Carl Schlichting considers that this is indeed a resurgence of interest in the study of plasticity after a period of about twenty years since Bradshaw, where it had been abandoned. Schlichting admits that he is unable to understand the reasons for this long pause and can only offer a few hypotheses:

> Until 1980, theoretical work on plasticity was limited. . . . The reasons for such neglect are puzzling, especially considering the clarity of Bradshaw's review. Surely part of the problem was the growing fascination with the detection and measurement of "genetic" variation, of which plasticity must have seemed the antithesis. Another problem was that environmentally induced variability in an experiment is typically avoided at all costs. Experimental complexity and the problem of measuring plastic responses also retarded progress. (Schlichting 1986, 669)

For Schlichting, the lack of interest in plasticity is linked to the need for reductionist experimentation to eliminate "confounding factors." What Schlichting fails to see in his analysis, however, is that when biologists took up the question of the evolution of plasticity in the 1980s, they did so mainly through a genocentric approach that did not really draw on the depth of plasticity as a concept. Moreover, like Bradshaw, they thought that the different "expressions" of plasticity (histogenetic plasticity, phylogenetic plasticity, genetic plasticity, to use de Beer's terms) all depended on the same kind of genetic control. So they naturally concentrated on what they knew: genotypic plasticity. The "neglect" that Schlichting mentions is only the result of a blindness associated with the consideration of "exclusively genetic" plasticity. But I also think that Schlichting is mistaken about the causes of this apparent neglect: they are not only the result of a focus on genetics, they are also linked to the polysemy of the term *plasticity*.

Indeed, if we look at all the different fields of biology during the same period, we see that the study of plasticity has always remained active. Between
1965 and 1980, many ecologists, evolutionists, and botanists (Harper 1967; Re-
hfeld 1979; Baker 1974) were interested in questions about plasticity of living
organisms both from a physiological and morphological point of view and
certainly not only from a genetic point of view. Their understanding of these
issues was based on the analysis of dynamic phenomena such as homeostasis (Ashby 1952; Thoday 1953, 1958; Hyde 1973) or canalization (Waddington
[1957] 2014, 1961). Even if all these studies did not explicitly and technically
refer to the notion of plasticity (although some did), it is clear that they started
from the idea that there is some plasticity in living things, as I will now show.

As an example of the interest in plasticity during this "hiatus," let's take a
closer look at Arthur Shapiro's (1976) article entitled "Seasonal Polyphenism."
In this article, Shapiro not only provides a detailed analysis of the phenomenon of phenotypic plasticity with a critical reference to Bradshaw's notion, but
he also carries out an exhaustive literature review, referring to other figures in
the history of plasticity such as Nilsson-Ehle, Schmalhausen, Goldschmidt,
and Waddington. Furthermore, Shapiro approaches the question of plasticity
from an empirical point of view, using an example: the fact that butterflies
can take on different colors within the same species depending on the time
of year in which they develop, a phenomenon known as seasonal polyphenism, a case study that will become central to later analyses of "developmental
plasticity." As I will discuss in chapter 8, this developmental plasticity is different from Gavin de Beer's histological plasticity; it refers to the phenotypic
plasticity due to developmental processes. Developmental plasticity will be
the focus of many studies from the 1980s in the field of evolutionary developmental biology (Nylin, Wickman, and Wiklund 1989; Nijhout 1991; Brakefield
et al. 1996). Of interest here is how Shapiro's article illustrates the vibrancy of
the debate in biology at that time around the phenomenon of plasticity.

Furthermore, while Bradshaw's definition influenced most studies on
plasticity developed between 1965 and the early 1980s, theoretical and conceptual views on the definition of *plasticity* continue to be debated long after
Bradshaw.[7] For instance, Shapiro's article distinguishes between how plasticity can be expressed differently in plants and animals:

> Polyphenism in plants [a result of phenotypic plasticity understood as a process]
> has fairly recently been reviewed in depth (Bradshaw 1965) and I will not pursue
> it here; the remainder of this paper applies to animals. The evolutionary problems
> in plants and animals are essentially the same, except as complicated by differ-
> ences in individual vagility and chromosomal flexibility. (Shapiro 1976, 262–63)[8]

While Bradshaw's work focused on plants, most of plasticity studies carried out by later biologists were on animals. Shapiro was trying to develop Bradshaw's view in this direction. Furthermore, Shapiro's study was part of a trend started by zoologists in the early twentieth century—and later continued by ecologists and ethologists—to define morphogenetic plasticity and its relationship to genotypic plasticity. It is also part of the emerging field of systems biology, which, instead of looking at isolated genetic information as responsible for a given phenotypic outcome, analyses the complex regulatory networks that need to be analyzed from a more general (systems) point of view. The late 1970s also saw the emergence of the first studies in sociobiology (a field that would gradually be integrated into modern synthesis), which focused on the role of plasticity in evolution (e.g., Cavalli-Sforza 1974).

In 1983 the biologist Stephen Stearns initiated a symposium, the proceedings of which were published in the *American Zoologist*, titled "The Inter-Face of Life-History Evolution, Whole-Organism Ontogeny and Quantitative Genetics" (Stearns 1983). The topics discussed and the participants in this symposium illustrate that biologists' understanding of plasticity was not limited to Bradshaw's or Schlichting's views. Plasticity has not been studied solely with the tools of quantitative genetics (although this discipline is gradually gaining ground in all studies of the evolution of plasticity). For instance, in Stearns's symposium, the ecologist Hal Caswell presented a series of reflections on the demographic effects and evolutionary consequences of phenotypic plasticity in life-history traits (Caswell 1983), and Sandra Smith-Gill discussed the issue of developmental plasticity, distinguishing two types of processes that she calls "developmental conversion" and "phenotypic modulation." Smith-Gill's paper will be a landmark in the history of plasticity: for the first time, a distinction is made between the fact that "organisms use specific environmental cues to activate alternative genetic programs controlling development" and that "nonspecific phenotypic variation results from environmental influences on rates or degrees of expression of the developmental program, but [without altering] genetic programs controlling development" (Smith-Gill 1983, 47). This distinction corresponds exactly to the distinction that will be made between a *developmental* and an *evolutionary conception* of plasticity (see chap. 8).

Contrary to Schlichting's claim, it seems that Sara Via and Russell Lande's model to explain the evolution of plasticity is not the result of a simple renewal of the studies on phenotypic plasticity but rather the direct consequence of a context in which the analysis of plasticity was lively and controversial. From the outset, Via and Lande's definition is more nuanced than Bradshaw's:[9]

> Environmental modification of the phenotype is common in the quantitative (polygenic) characters of organisms that inhabit heterogeneous environments. The profile of phenotypes produced by a genotype across environments is the "norm of reaction" (Schmalhausen 1949); the extent to which the environment modifies the phenotype is termed "phenotypic plasticity" (Gause 1947, Bradshaw 1965). Because phenotypic response to environmental change may facilitate the exploitation of some environments and provide protection from others, the level of plasticity in a given trait is thought to be molded by selection (Gause 1947, Schmalhausen 1949, Bradshaw 1965), (Via and Lande 1985, 505)

The authors are obviously mindful of the conceptual controversies that still existed in 1985 and in particular insist on what distinguishes plasticity from the norm of reaction, where the genetic conception of plasticity had tended to erase this distinction by suggesting that the norm of reaction was the result of different phenotypic expressions for a genotype in different environments. The "genetic view" would not take into account the fact that plasticity could be the result of something other than differential genetic expression (expressed through the norm of reaction).

In their article, Via and Lande show how a new model can raise new questions about plasticity. They show, for example, that it is possible to measure the evolution of plasticity for a given species in environments that vary (e.g., the *Pontia* butterfly has more or less wing pigmentation depending on the season in which it develops), and that it is also possible to measure intergenerational plasticity when successive generations are exposed to fluctuating environments (e.g., the effects of global climate change on certain plants). In order to distinguish between these two types of situations, the notions of "labile" and "nonlabile" traits are adopted from Schmalhausen. Labile traits refer to the fact that the individual adjusts its phenotypic expression throughout its life (e.g., a plant will adapt to the amount of water available in the environment throughout its life), whereas nonlabile traits indicate that the expression of traits is fixed once and for all during development (e.g., some root cells of plants that have the function of absorbing water from the environment). In the first case, we observe that the norm of reaction evolves toward an optimum (e.g., the plant will get used to the average amount of water in its milieu in order optimize its growth). In the second case, the situation is much more complex: the equilibrium reached depends on the intensity and duration of the environmental fluctuations to which the populations are subjected (e.g., to obtain a root cell with hairs capable of absorbing water optimally). In 1989 Kirkpatrick and Heckman, followed in 1992 by Gomulkiewicz and Kirkpatrick, proposed quantitative genetic models to illustrate the evolution of the norm of reaction in these two types of situations.

But let's go back to Via and Lande and to the formulation of their first model in 1985. Based on their model, Via went on to propose a theory of the definition and the evolution of phenotypic plasticity. She agreed with population geneticists (e.g., Schlichting, Stearns, and others) that adaptive plasticity can evolve as a result of natural selection. However, there is disagreement on how to describe the evolution of plasticity. According to Via, "The assertion that phenotypic plasticity is a character that is independent of trait means and the attendant implication that plasticity itself is the target of selection" (Via 1993, 352) is based on Bradshaw's flawed analysis. This flawed analysis is attributed to the lack of sufficient quantitative genetic data in 1965. Via, for her part, drew two main conclusions from her observations from genetic studies (Via 1984a, 1984b): first, selection acts only on the phenotypic traits that are expressed at a given time and in the given environment where the individual is located, and second, within each environment selection acts to move the population toward an optimum phenotype. It follows logically that the evolution of adaptive norms of reaction can only occur through the phenotypic traits themselves. Selection would not act directly on plasticity, which could therefore not be considered as a distinct trait with its own genetic etiology. However, the relationship between the evolution of plasticity and the evolution of the trait expressed in the environment remains in question.

So the difficulty is not resolved at all. A year after Via, in a paper published in 1986, Schlichting took a very different position. By comparing two species of purslane (*Portulaca grandiflora* and *Portulaca oleracea*), he claimed to show that the plasticity of a trait can evolve independently of the trait itself. Schlichting takes as an example the quantitative trait "growth of the plant," which is identified by the importance of root development. He notes that even if the mean of the trait is the same for different environments in the two species considered, the degree and direction of the plastic response can be different. There should be a trait that specifically controls this plastic response independent of the "growth of the plant" trait. Schlichting then concludes that the genetic control of plasticity can only be distinguished from the genetic control of the trait because the evolution of a trait is explained by its genetic basis, and because plasticity itself is a trait, the evolution of plasticity must be explained by its genetic basis. Peter Van Tienderen and Koelewijn, in 1994, and Gerdien de Jong, in 1995, would establish quantitative models to highlight the variation in plasticity according to Schlichting's model. These models confirm the existence of "genes of plasticity"[10] by empirically demonstrating the independence between the evolution of the trait mean and the evolution of plasticity.

In fact, the Via-Schlichting controversy illustrates the persistence of conceptual disputes rather than the existence of real empirical obstacles to identifying a "genetic" control of plasticity. While for some, a genetic control of plasticity does seem to exist, the problem remains as to what exactly is meant by "genetic control." This problem seems to be based more generally on the vagueness of the different conceptual approaches to plasticity—sometimes understood as any phenotypic trait that would be subject to natural selection and sometimes as a general property of living organisms whose evolution should be analyzed without knowing its precise causes.

In conclusion, it is clear that between 1965 and 1980 a number of biologists were aware of the conceptual complexity surrounding the use of the term *plasticity* in biology and attempted to explain it theoretically (although they continued to rely methodologically on the tools of quantitative genetics).

DEFINING GENES OF PLASTICITY?

In line with the genetic view of plasticity, and in the context of research in quantitative genetics, Scheiner and Lyman in 1991 established a classification of what they called the "genetic bases of plasticity."[11] According to them, three distinct categories cover the genetic basis of the plastic response. First, "overdominance"[12] expresses the fact that plasticity is an "accident" because both parents expressed homogenous phenotypes because of homozygous genotypes, and the resulting heterozygous genotype leads to a "plastic" phenotype (as being outside the range of the parents' phenotypes) (Lerner 1954; Gillespie and Turelli 1989). Second, "pleiotropy"[13] expresses the fact that plasticity is a function of the differential expression of the same gene (the same set of alleles) in different environments (Falconer [1960] 1981; Via and Lande 1985, 1987; Via 1987). Finally, "epistasis"[14] expresses the fact that two classes of genes control the two basic characteristics of a norm of reaction: its plasticity and its overall mean. The degree of plasticity, in this view, is then due to the interaction between the genes that determine the magnitude of the response to environmental influences and the genes that determine the mean expression of the trait (Lynch and Gabriel 1987; Jinks and Pooni 1988; Scheiner and Lyman 1989).

This model, unlike Via's, assumes that the trait mean and the environmental variance are two independent factors. For Scheiner and Lyman (1991), these three categories are not mutually exclusive, and their use applies to effects that occur not in a single environment but in different environments over time. This approach is essentially phenomenological, that is, based on observing "types" rather than investigating the actual causes of plasticity. However, the underlying assumption is that these different types can all be captured using the same tools, those of genetics. From this perspective, statistical studies of quantitative genetics should be sufficient to study models in general without the need to know what the actual role of genes is (De Jong 1995). The good functioning of the model through its predictive capacity becomes a sufficient proof of its validity.

The position defended by Sara Via, although made difficult by the "polynomial approach" (that of Van Tienderen 1991; Scheiner 1993; Van Tienderen and Koelewijn 1994), is not rejected (see Via et al. 1995). For Via, the "presumed" independence between the "trait mean" (the mean phenotypic expression of the trait) and plasticity, defended by the proponent of the polynomial approach, remains to be confirmed. Via shows that the trait mean can be measured both for a single environment (environment-independent variability) and from a possible range of trait expression, this time reflecting variation in the environments in which the trait would be expressed. In the latter case, Via calls the trait mean the "grand mean." She suggests that the distinction between these two measures (trait mean and grand mean) allows the question of the correlation between the evolution of the trait and the evolution of plasticity to be revisited. For example, in the case of the trait "plant growth," the grand mean may be similar in two different species (comparatively, both species will grow the same amount), whereas the trait mean will be different for each species in a single environment (one species will grow more than the other in the E1 environment and vice versa in the E2 environment). This means that plants of different species will grow differently in the same environment, but overall, their mean growth will be the same when these different species are exposed to changing environments. Via defends the idea that phenotypic plasticity is not a specific trait but a *by-product* of the selection of different means of the phenotypic trait in different environments. In her view, the reality is therefore more complex than it appears, because there will necessarily be an interaction, albeit indirect, between the two variables.

In order to resolve this controversy, which they consider to be the result of a "semantic wrangling" rather than the expression of a fundamental divergence between two positions—that of Via and that of the polynomial approach (defended in particular by Scheiner)—Carl Schlichting and Massimo Pigliucci proposed in 1993 a definition of "plasticity genes" in terms of "regulatory loci that exert environmentally dependent control over the expression of structural genes and thus produce a plastic response." For both authors, the alternative between "the existence of plasticity genes" and "plasticity as a by-product of selection" is only apparent: the two possibilities are not mutually exclusive. Moreover, the evidence for the existence of plasticity genes in the recent literature of the last decade seems to confirm their hypothesis, as shown by the paradigmatic example of the genes encoding light-sensitive phytochromes in plants. This last example, described by Schmitt, McCormac, and Smith in 1995, is now one of the main arguments in favor of the existence of genes coding for plasticity (at least in terms of growth).[15]

MOLECULAR CONTROL OF PLASTICITY?

In 1996 Pigliucci revisited his 1993 definition of "plasticity genes" and restricted it to the idea of "regulatory loci that directly respond to a specific environmental stimulus by triggering a specific series of morphogenic changes." It should be noted that this definition does not imply that all regulatory genes are plasticity genes: indeed, not all regulatory genes respond in the same way to environmental stimuli (see Pigliucci 2001). The genetic basis of any plastic response will therefore necessarily involve many more genes than those directly linked to environmental sensing. However, the identification of the latter category of genes (those directly related to environmental sensing) is conceptually important because their existence cannot be explained without recourse to the action of natural selection. It is therefore likely to attract increasing attention from biologists interested in the molecular basis of plasticity.

In 1990 Harry Smith also explored this molecular pathway in a special issue of the journal *Plants, Cells & Environment* on "sensing the environment." He investigated the nature of the molecular mechanisms that link the perception of environmental signals to specific developmental responses (which, according to Pigliucci's 1996 definition, correspond to phenotypic plasticity). From this study, Smith concluded that it is the *differential regulation* of the expression of members of multigenic families that provides the molecular basis for phenotypic plasticity.

The first molecular studies of genotype-environment interactions show the existence of specific responses induced by a particular type of stress, responses induced by a limited number of constraints, and generalized responses to a variety of stressful situations. At the same time, epistasis and pleiotropy at the molecular level are receiving renewed attention, making it increasingly difficult to interpret plastic response patterns in the absence of molecular information. The concepts of epistasis and pleiotropy were introduced in the mid-1990s at the molecular level with a different meaning from

that used in quantitative genetics.[16] These new concepts led to the abandonment of old definitions of plasticity in favor of new definitions centered on
regulatory mechanisms at the molecular level.

From these various studies carried out at the molecular level, one observation can be made: the expression "plasticity genes" is being used less and
less. This is not because there is no evidence for a genetic control of plasticity
(on the contrary, the various studies carried out all seem to confirm its existence) but rather because the question of the direct or indirect link between
genes and traits has gradually been abandoned in favor of questions about
the *proximate causes* of plasticity, mostly understood as genetic or epigenetic
regulators (e.g., Belsky et al. 2009). In other words, the question has shifted,
and the main issue facing biologists can now be summarized with the following question: Is plasticity characterized by simple allelic sensitivity, or is
it controlled by genes or more complex gene functions that play a regulatory
role on the genes on which the trait depends?

This change of perspective and this shift in the problem, correlated with
the expansion of this new field of investigation—molecular biology—allows
us to develop the idea, originally proposed by Woltereck and taken up by
Schmalhausen and Waddington, according to which the norm of reaction
is transmitted and can evolve. This is also a response to the growing empirical evidence that there is no direct causal link between a genotype and a
phenotype.[17] Rather, the phenotype appears to be the product of a complex
epigenetic system that includes both genes capable of interacting with internal and external signals and genes capable of producing those same signals.
It is these complex epigenetic systems (rather than specific genetic or allelic
variations) that are passed on through evolution.[18] From this perspective, a
significant amount of work in molecular biology and physiology has directly
addressed the molecular basis of phenotypic plasticity (e.g., H. Smith 1990;
Callahan, Pigliucci, and Schlichting 1997; Aubin-Horth and Renn 2009). This
work, which originally focused on the genetic basis of plasticity, is no longer
limited to genes alone. The functionally flexible hormonal systems of plants
and animals provide a starting point for understanding how environmental
signals are translated and interpreted and how organisms respond to them
(e.g., Friml and Sauer 2008). Indeed, hormones appear to be the main interface between the genetic level of action and the external environment in
the sense that they perform two important functions: they help to shape the
organism, and they transmit signals about environmental states from sensory
receptors to cells, triggering specific reactions that characterize phenotypic
plasticity. For example, the biologist Frederick Nijhout showed in 2003 that
the development of alternative phenotypes (both in reaction norms and in

polyphenisms—both continuous and discrete) may be due to specifically evolved mechanisms that are themselves regulated by variations in the pattern of hormone secretion (Nijhout 2003). Alexander Badyaev showed in 2005 that the phenotypic assimilation of the stress response is facilitated by the common involvement of neural and endocrine stress response pathways in other functions of organisms. Finally, Erica Crespi and Robert Denver observed in 2005 that the neuroendocrine stress axis represents an ancient phylogenetic signaling system that allows the fetus or larva to adapt its development rate to prevailing environmental conditions.

These various works illustrate the fact that, although empirical evidence for the existence of a genetic basis for plasticity has been accumulating since the mid-1980s, the upheaval in the understanding of the molecular mechanisms—and no longer just genetic mechanisms—at the origin of traits has led biologists to question once again the definition of plasticity. The problem associated with the evolution of plasticity, that is, the attempt to understand plasticity as explained by natural selection, does not seem to be able to be resolved without first going through a theoretical and coherent definition of this notion, which has emerged decade after decade.

Thanks to genetics, plasticity has acquired a definition and a central place among the "operational" concepts of biology. This acquired status is linked, on the one hand, to Bradshaw's seminal work on phenotypic plasticity and, on the other hand, to the renewed interest in the field of research linked to biological plasticity, first in ecology from the 1970s and then, from the 1980s, with the development of new tools in quantitative genetics. However, the emergence of new tools does not make it possible to "refine" the definition of the concept of plasticity, which, unlike other concepts, such as "epistasis" or "pleiotropy," remains stubbornly controversial among biologists. A number of important issues arise concerning the different scales of analysis of plasticity (depending on whether plasticity is considered within a generation or between generations), the genetic determinism of plasticity (the existence of plasticity genes, plasticity as a by-product of evolution, etc.), and, finally, the definition of phenotypic plasticity and its possible distinction from the reaction norm. In any case, as a result of these reflections on the evolution of plasticity, it is becoming increasingly difficult to give a "general" definition of plasticity in biology, just as it is becoming very difficult to separate this concept from the field of genetics.

Despite the lack of a definition that all biologists would agree on, and despite the fact that understanding the evolution of plasticity would require such a definition, biologists continue to refer to the term. Two conclusions can be drawn from this double observation. The first is that biologists, even if they agree that there is a certain polysemy of the term *plasticity* in biology, nevertheless run the risk of confusing "apples and oranges" by continuing to refer to the term without necessarily defining it. Massimo Pigliucci, in his 2001 book, reminds us of the existence of "different types" of plasticity and shows how confusion can arise (Pigliucci 2001, 42).

But Pigliucci also seems to assume that all the meanings of plasticity are somehow connected and that we should be aware of these "deep biological correspondences." However, the existence of connections between the different uses of plasticity does not mean that there are not important conceptual distinctions within the uses of the term. In other words, there may not be a single ontological meaning behind plasticity with different biological manifestations. Moreover, it seems that there are two basic ontological meanings of plasticity: either plasticity is an explanans of variation, or it is explained from the point of view of evolution (an explanandum of evolution).

These two meanings have consequences for the role of plasticity in evolutionary thought. Darwin used it to fill a gap in his knowledge of evolutionary phenomena because he was unable to provide a satisfactory explanation for the laws of variation. The notion of plasticity appeared on the margin of scientific explanation.

Before him, embryologists had used plastic- terms to assess a predisposition of developing tissues, layers, and cells to produce other kinds of tissues, layers, and cells. They were somehow interested in how variation was produced diachronically. After Darwin, the first geneticists, in line with Mendel's mentorship, looked at how variation is produced synchronically when they compared how parents could produce offspring with trait variations for a given character. They used plastic- terms to refer to the variability "produced" by a single genotype.

Modern evolutionary biologists have attempted to provide a synthesis of these views to fill Darwin's gap, symbolized by his use of the term *plasticity* as a general organizing principle. However, their focus was mainly on the laws of heredity rather than the laws of variation. By focusing on the "inheritance" of geneticists, the ontological meaning of plasticity in relation to the diachronic production of variation was somehow erased from synthetic evolutionary thinking. As a result, today the notion "phenotypic plasticity" (the "phenotypic" attribute of which illustrates its link with genetic terminology) refers to the idea that plastic phenomena are to be explained by neo-Darwinian mechanisms of natural selection (and thus phenotypic plasticity appears as an explanandum of evolution). It remains to be explained what the biological phenomena are under this precise ontological meaning of plasticity (I explore this in chap. 7).

However, as this chapter has shown, there is another ontological meaning of plasticity that has yet to be clearly explored (although there may be biological link between the two): that of plasticity as an explanans of variation. In the next chapter, I examine how developmental biology today provides good examples of this meaning of plasticity.

In the previous chapter I showed how evolutionary thinking, by bringing together questions about variation and heredity, highlights the fact that the notion of plasticity in biology can appear as a boundary concept not only because it is used in different fields of biology to showing connections between different biological phenomena but also because it helps to bridge the gap between different epistemologies inherent in the fields of study. However, until the late 1980s, because evolutionary biology was dominated by the attempt to understand heredity, the concept of plasticity in evolution was understood in similar way to its use in genetics.

In the next chapter I show how the rise of developmental biology brought back a different meaning of plasticity (the active sense of embryology and the focus on plasticity as an explanans of variation), a change that would be crucial for the contemporary understanding of plasticity in evolutionary biology. Gavin de Beer's hope was that "progress in knowledge of [genetic plasticity would] increase our knowledge of [evolutionary plasticity]." I have shown that genetics has increased our knowledge of plasticity, in particular, by looking at the genetic basis of the phenomena. But geneticists, and later evolutionary biologists of modern synthesis, focused on a very specific meaning of plasticity—"plasticity" understood in a passive way and as an explanandum of evolution. As a result, de Beer's whole synthetic view of plasticity in evolution, including the idea that "the variability that plasticity denotes must be studied from the embryological aspect"—has not been studied as much from an evolutionary perspective. Some would argue that a synthesis in evolution does not need this second aspect, but if we take it seriously, the elucidation of "plasticity" (understood in an active way and as an explanans in evolution) could help to increase our knowledge of evolutionary processes.

PART 2

Biological Plasticity:
Two Ways to Explain Variation

6

Biological Plasticity, an Explanans in
Developmental Process

At the beginning of its journey, development is plastic, and a cell can become many
fates. However, as development proceeds, certain decisions cannot be reversed

CONRAD WADDINGTON ([1957] 2014)

While the association of plasticity with development is straightforward—
developmental entities would be plastic—the reasons for such an association
are not always straightforward. The purpose of this chapter is to clarify such
an assumption. Two of the most commonly used terms to explain plasticity in
development, *embryonic induction* and *developmental regulation*, have been
used to explain variation within developmental processes. In the previous
chapter we saw that plasticity in biology can be understood in two different ways that remain incommensurable—as an explanans of variation in a
developmental view or as an explanandum in an evolutionary context. This
chapter examines plasticity as an explanans of variation through its association with embryonic induction, developmental regulation, and developmental processes. In other words, I address the question, What does plasticity
explain about variation?

The explanans and the explanandum are concepts borrowed from the
general philosophy of science.[1] Since the explanans is mainly understood in
relation to its respondent, the explanandum—what is explained—this chapter introduces some elements that will also be useful for the next chapter,
which is devoted to plasticity understood as an explanandum in evolutionary
biology. The first part of this chapter is a careful examination of these debates.

Returning to the main theme of this chapter, plasticity in developmental
biology, I would say that any observed variation is posited as the result of
some developmental or physiological change. The observed variation appears
at a tissue/structural level, a cellular level, or a molecular level. If the cellular
level is the scale at which we are looking, then the question remains as to
how a single cell, or a small number of different cells, is able to generate a
huge diversity of cell types. Another way of framing the same question would

be similar to the fundamental goal Norman Wessells sets for developmental biology in his book *Tissue Interactions and Development*: "How [are] various types of information (information from genes, cytoplasmic substances, relative positions, physiological and chemical gradients, the environment) . . . integrated to allow a spherical egg a fraction of a millimeter in diameter to become a starfish, a butterfly or a human being?" (Wessells 1977, 15).

Behind this question lies another. Since development can be seen as a relatively stable and constrained process, to what extent is it able to generate novelty out of this regularity? It seems that it is at this level that plasticity might play a role in "explaining" variation despite the regularity of the process. However, while novelty seems to be a good way of looking at the question of generation of variation, the role of the developmental process in explaining novelty remains highly controversial, even among developmental biologists. For example, the biologist and historian Brian Hall argues that novelty results from "tinkering" with existing processes, the consequence of modifications to existing gene networks that also involve the selective advantage provided by gene duplication or modification (Hall 2005). An example of this type of reorganization of genetic "building blocks" to generate new traits is provided by the theoretical model (fig. 6.1), showing how the body plans of insects (in this case the grasshopper) and crustaceans (the planktonic shrimp) could have diverged from a hypothetical common ancestor based on the differential expression of homeotic genes (genes that regulate the development of anatomical structure).

But such a view already involves complex machinery. To "tinker" you need "tools." This is possible in complex higher organisms. What about the emergence of simple but essential changes such as cell polarization (the fact that cells have asymmetric gradients of molecules that polarize them from very early stages of development) or multicellularity? Is it also originally based on the same type of complex genetic mechanisms? For example, before explaining the appearance of the different forms of horns in deer, it is first necessary to explain how the single egg cell that gives rise to these deer was able to divide and differentiate to form a complexly organized living being. Norman Wessells raised this question when he said about the very early stages of ontogenesis, "Cells elongate, become narrow, or move about, and so generate the changing form of an embryo" (Wessells 1977, 99). Differentiation no longer seems to be a problem associated with multicellularity, but a single cell alone can undergo a process of differentiation. As a result, variation appears to be a fundamental property of living systems and not limited to complex systems. The problem for biologists here is similar to the one that Caspar F. Wolff faced in 1759 when he described the development of the chick's heart: it was already

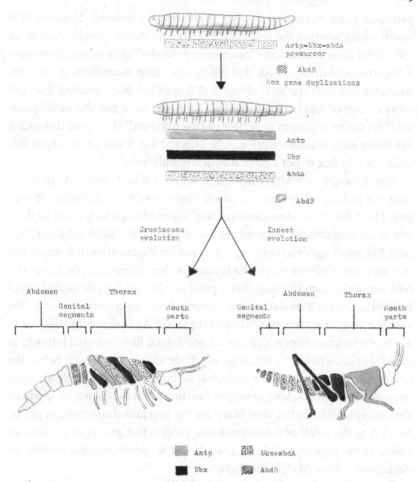

FIGURE 6.1. Theoretical model used to observe how the body plans of insects and crustaceans might have diverged from a hypothetical common ancestor depending on the differential expression of Hox genes. The comparison is between a planktonic shrimp (crustacean) and a grasshopper (insect). Drawings by Anton© based on Gilbert (2006).

a question of understanding how an avian embryo could develop through progressive variations into its "mature" (posthatching) form.

In discussing variation, it is important to distinguish between the specific problem of ontogenetic or individual development, which is concerned with gradual and rapid changes in form, and other problems—found in evolutionary biology and often explained by statistical models, such as the distribution and changes in frequency of versions of a gene (or allele) in populations (slower than ontogeny), by population genetics. To address the specific question of the "generation of variation" during ontogeny from an evolutionary perspective, the notion of "novelty" has been put forward. However, it is worth asking whether the notion of "novelty" is entirely appropriate here. In 1960 Ernst Mayr first defined "evolutionary novelty" as "a newly arisen character, structural or otherwise, that differs more than quantitatively, from the character that gives rise to it" (Mayr [1960] 1997, 89). Mayr clarified that "not every change of the phenotype qualifies" and that those that did would probably "permit an organism to perform a new function." He argued that such a definition must remain tentative, as he believed that it was often "impossible to decide whether or not a given function is truly 'new.'"

The difficulty with Mayr's definition is that it is both restrictive in that it does not include quantitative variation (since novelty, by definition, is supposed to differ "more than quantitatively"), something which could be decisive in some situations, especially at the cellular level. (Small cells and large cells, like small eggs and large eggs, may not have qualitative differences, but they may have different properties because of their difference, which could be seen as novelty from an evolutionary point of view; see, e.g., Newman [2011].) In addition, its limits remain vague—novelty becomes synonymous with "the appearance of a new function." Despite the fact that it is sometimes difficult to say how a function is new, as Mayr acknowledges, function itself is linked to the idea of evolutionary adaptation, which somehow puts the cart before the horse, since novelty should by definition be fitness neutral. Largely for these reasons, I argue that linking novelty to function is not a good way to approach the concept, and a better way, based on my previous distinction, might be to see it as the result of a developmental process that produces a difference visible at the populational level. I would call it "developmental novelty" to distinguish it from Mayr's definition.

In fact, the main problem we face in dealing with developmental novelty is not so much how to integrate causal-mechanical explanations (concerning the appearance of this novelty) and statistical or evolutionary explanations (concerning the adaptive character of this novelty). Rather, it is the problem of how to explain the appearance of a particular new form in a unified way from different causal-mechanical explanations—given that the developmental novelty would result from different and distinct causal mechanisms. From this perspective, the integrative approach to explaining a developmental novelty also involves "grasping" in a unified way the different types of causal-mechanical explanations that give rise to that developmental novelty. An illustration of this type of explanatory integration is provided by a model (fig. 6.2) described by developmental biologist Stuart Newman (2010) in the book *Evolution: The Extended Synthesis*.

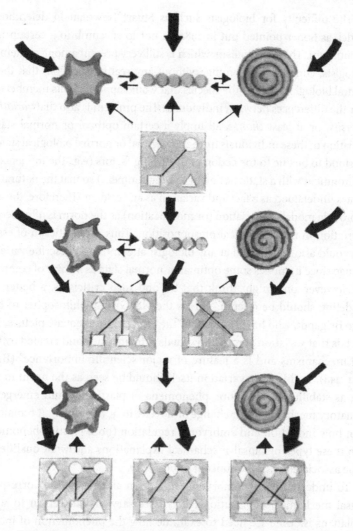

FIGURE 6.2. Diagrams showing three different morphotypes: "star," "worm," and "snail." The theoretical model shows the evolutionary partitioning of a morphologically plastic ancestral organism into distinct morphotypes with unique genotypes. (*Top*) Hypothetical primitive Precambrian metazoan that can change shape despite having a single genotype (represented by the box with symbols). (*Middle*) Descendants of the top organism after an episode of stabilizing evolution. (*Bottom*) Modern organisms descended from the organisms in the middle. Advanced stabilizing evolution has now led to each morphotype being associated with a unique genotype. Newman used arrows to represent the different causal-mechanical explanations at the different "stages" of evolution. The relative importance of each of these causal-mechanical explanations (environmental, molecular, retro-regulatory) over the evolutionary time scale is indicated by the relative size of each of these arrows. Drawings by Anton© based on Newman (2011).

The difficulty for biologists such as Stuart Newman in defending their model, as Sober pointed out in 1980, is not to succumb to a certain form of essentialism. This essentialism, which is still very present among contemporary biologists, consists in assuming, without any real justification, that there is a natural biological state of the species that would apply to all its members, whatever the differences between individuals. The problem is also that *variation* and *diversity*, or at least *change*, all imply a certain optimal or normal state. The deviation of these individuals from this optimal or normal biological state is understood to be due to the action of disturbing factors (e.g., the influence of the environment with a static notion of the environment) so that the natural state is always understood as a fact and variation as an accident. Therefore, the attempt to provide models of variation (or modification) as the norm is often seen as an aberration in this theoretical presupposition. Thus, this criticism of essentialism could also be directed at any biologist attempting to describe variation or change since it implies some optimal or normal state as a point of comparison.

However, one might think that beyond such criticisms, a better way of modeling should be offered. This is the view of the philosopher of biology Ingo Brigandt, who has argued that "the empirically adequate picture, in contrast, is that variation across individuals is maintained and created by biological mechanisms and is a feature of major scientific importance" (Brigandt 2011, 258). In this view, variation itself should be seen as the norm in nature, just as stability is. Therefore, phenomena of plasticity could emerge as explanatory mechanisms for change or novelty in development. It remains to be seen how induction and embryonic regulation (both plastic phenomena) fit into these types of causal-mechanical explanations and what qualifies them to be associated with the concept of plasticity.

To understand how embryonic induction and regulation correspond to causal-mechanical explanations, it is necessary first to return to what an explanans is. Then we need to examine how the phenomenon of induction explains the generation of modifications during development. Finally, I conclude the chapter with my understanding of the concept of plasticity as an explanans of variation in the developmental process.

Explanans in Biology: Explaining Variation

Carl Hempel, in his Philosophy of Natural Science (1966), shows that if man has persisted in trying to make sense of extraordinarily diverse, often puzzling and sometimes threatening events that occur in the world, he has long done so by referring to myths and metaphors of his own making. And in a sense, we could argue that all models are metaphorical or iconic. "Some of these explanatory ideas are based on anthropomorphic conceptions of the forces of nature, others invoke hidden powers or agents, still others refer to God's inscrutable plans or to fate" (Hempel 1966 47). According to Hempel, in order to provide a true explanation, scientific models must meet two systematic requirements: "explanatory relevance" and "testability" (48).

In order to provide scientific explanations that satisfy both requirements, Hempel proposes the model of the deductive nomological (DN) explanation, which links, in the form of an inference, general propositions (general laws) and particular or singular propositions (descriptions of proven facts), which together form the premises or explanans of the explanatory reasoning and, on the other hand, the phenomenon to be explained by an explanation—the explanandum—which leads to the necessary conclusion of the argument.

If we try to apply this model to our subject, the question might be, What might be the explanans of variation in biology? According to the DN model, variation is what needs to be explained within the framework of evolutionary theory. Indeed, the explanans is supposed to be characterized both by a nomological statement (i.e., governed by a general law) and by a particular statement related to the biological context that one is trying to explain. Most philosophical debates about biology have focused on the nomological part of the explanans. And in biology this question is particularly sensitive because of the absence of strictly universal laws. Darwin did title chapter 5 of *On the Origin of Species* "Laws of Variation," but he ended the chapter on a defeatist note: "Our ignorance of the laws of variation is profound" (Darwin 1859, 167). Unfortunately, the situation does not seem to have changed much since Darwin: it remains difficult to identify laws of variation (especially regarding changes during development) that are similar to the general laws of physics, and indeed no one seems to be really looking for such laws anymore. But the problem remains. Does this mean that we are still unable to provide a satisfactory scientific explanation in biology? In recent years, many biologists have tried to answer this aporia, both through genetic analysis and through the study of development and its regularities, although this does not seem possible within the framework of a DN statement. The idea is to identify strict regularities that have the force of laws.

In order to properly consider the question at hand, it is appropriate to first look at the problematic part of the explanans in biology: how to treat laws in biology. Hempel himself was well aware that not all scientific explanations are based on laws of a strictly universal form. The difficulty lies in being able to distinguish between regularities that have the value of laws and those that are purely contingent or random. In order to give a scientific account of a large number of phenomena, while at the same time satisfying his criteria of scientificity, Hempel proposed a number of alternative models, such as the statistical inductive (SI) and statistical deductive (SD) models, which are based on probabilistic explanations. In both models, the explanans is no longer characterized by universal laws but by probabilistic laws such as "There is a high probability that people in contact with a person with Covid-19 will also contract the disease." In this respect, Darwin had already seemed to warn that any kind of probabilistic approach could not be applied to variation: "Not in one case out of a hundred can we pretend to assign any reason why this or that part [should] differ, more or less, from the same part in the parents" (Darwin 1959, 167). It seems that in the context of biology, and in particular with regard to the explanation of variation, the theoretical framework of the explanatory model proposed by Hempel is problematic, at least with regard to the nomological part of the explanans.

It has since been shown that the criteria of the DN model provide an unsatisfactory scientific explanation. In 1971 Wesley Salmon proposed the following now famous example to illustrate the inadequacy of the DN model:

"(L) All males who takes birth control pills regularly fail to get pregnant;
(K) John Jones is a male who has been taking birth control pills regularly;
(E) John Jones fails to get pregnant." (Salmon 1971, 32, quoted in Woodward and Ross 2021)

This explanation is deductively correct according to Hempel's model and is based on a perfectly correct law. However, it cannot reasonably be regarded as a satisfactory scientific explanation for the fact that John Jones did not get pregnant. To get around this sort of difficulty, Salmon proposed to introduce a form of asymmetry into the laws by insisting on the temporal priority of the explanans over the explanandum and by focusing on causal laws, thereby suggesting that the main problem with the previous false explanation was that it was a *noncausal* explanation, that is, that the cause of John Jones's failure to be pregnant was due to his lack of female reproductive organs, not to his taking birth control pills.

A few years later, Philip Kitcher took up Salmon's example and proposed an alternative view of explanation, this time tending toward a "unification of knowledge" (Kitcher 1989, 485). According to Kitcher's approach, an explanation is not a logical deduction from a law or a general proposition with the value of a law (as in the DN or DS model) but a deduction that follows from a *unified system of our global knowledge*. The inference about John and the birth control pill is no longer an appropriate explanation in this framework because there is a broader inference that leads to the fact that no man gets pregnant and explains why. Kitcher thus relies on a much more complex scientific explanatory structure than that presented by his predecessors. In particular, he shows that a number of "schematic sentences" can be identified. For example, the sentence "Organisms homozygous for the sickle cell allele develop sickle cell anemia" can be replaced by the schematic sentence "Organisms homozygous for A develop P." To these schematic sentences can be added a number of "filling instructions," which are rules for specifying how to replace letters in schematic sentences. This is how formal logic works. For example, these instructions might tell us to replace A with the name of an allele and P with the name of a phenotypic trait. Then the "schematic arguments" are sequences of schematic sentences. The "classifications" describe which sentences in the schematic argument are the premises, which are the conclusions, and finally which laws of inference are used. Finally, the "argument pattern" is an ordered triplet consisting of a schematic argument, a set of filling instructions (one for each of the terms in the schematic argument), and a classification of the schematic argument. Within this general scheme, which is extremely complex, Kitcher will show that the more an argumentative structure imposes restrictions on the arguments that instantiate it, the more stringent it should be regarded as being: "Science advances our understanding of nature by showing us how to derive descriptions of many phenomena, using the same pattern of derivation again and again, and in demonstrating this, it teaches us how to reduce the number of facts we have to accept as ultimate" (Kitcher 1989, 423). For Kitcher, it is this progressive restriction that allows us to unify the field of our knowledge.

For some years now, many philosophers of science have rejected Hempel's model on the ground that it is not the only valid model of scientific explanation. On the other hand, Kitcher's model sometimes seemed simply not to be useful when all we have is statistical or probabilistic data. Moreover, some philosophers have tried to show that a single justification for explanation is insufficient and that different models can be proposed. This idea has been pursued particularly in the philosophy of biology. Some have argued that biology can aim to explain individual facts and some generalities without referring to laws of nature (Brigandt and Love 2008). The ideal of logical positivism implied reference to formal notions to account for a good explanation, and the DN model was intended to provide a syntactic approach to statements that capture the laws of nature. Such an attempt has been unsuccessful in biology. As Brigandt said, "In contrast, what makes a particular explanation in biology scientifically acceptable is likely to depend on empirical considerations that cannot be characterized in a formal-syntactic fashion" (Brigand 2011, 258). In other words, causal-mechanical explanations (see Schaffner 2006; Morange

2012) may be more adequate than formal logical models. Thus, and in light of these theoretical considerations, many philosophers of biology are tempted to endorse a certain form of epistemological pluralism regarding how to proceed to scientific explanation in biology (Brigandt 2013, 76).

The idea behind the pluralist epistemological position is that biological explanations can take different forms in different situations and that what is called an "adequate explanation" depends mainly on pragmatic empirical considerations. Some philosophers of biology have gone so far as to use this pluralist view to take very bold antireductionist positions. For example, they have argued for an ontological pluralism, a "disunity" of biology (Dupré 1993; Rosenberg 1994). The disunity described by these authors allows us, through the plurality of existing projects and perspectives, to account for the existence of different, separate, and isolated fields that are the site of true explanations. Dupré gives various examples of questions—molecular phenomena occurring during the development of individuals, the transmission of phenotypic traits, or the evolution of genetic predispositions of populations—that are in one way or another related to genetics (Dupré 1993, 95). However, he points out that despite the (unified) development of molecular genetics, all these issues cannot be treated in a unified way. On the contrary, it seems that the explanations available can only be specialized according to the field of study concerned (developmental biology, ecology, population genetics, etc.).

In 2008, Ingo Brigandt and Alan Love rejected this "antireductionist" view and suggested that a distinction should be made between "theoretical reductionism" (or ontological reductionism)—which encompasses most philosophical discussions of explanation in biology but which they argue is really about *where* explanation should take place—and what they called "explanatory reduction" (which is about *how* explanation takes place). They argued that explanatory reduction does not necessarily take the form of a logical deduction from a theory containing laws, that is, the explanation of a complex whole by its parts. Thus "what explains," the explanans of the explanation, does not necessarily contain terms referring to entities of a "lower" level than to that of what is being explained (Weber 2004). On the contrary, and as Kenneth Schaffner showed in 2006, the explanans of an explanation may be a "fragmentary patchy explanation"; that is, in order to explain a phenomenon, it is sometimes necessary to go through different elements of different theories that may have succeeded each other over time (giving this "patchy" picture of the explanation).

Schaffner shows that the relationship between explanans and explanandum is not necessarily one of deduction of a theory from a more fundamental theory. He uses an example from physics to illustrate what he means by fragmentary patchy explanation. For example, to explain the dispersion of light as it passes through different media, it is necessary not only to rely on Maxwell's equations but also to take into account Lorentz's theory of electrons. But this will not be enough to explain all the physical phenomena of optics, since to explain a photoelectric effect it will also be necessary to appeal to the principles of elementary quantum mechanics, and to explain the optics of moving bodies it will also be necessary to refer to Einstein's special theory of relativity. Thus, in the case presented, it is not the "theoretical reduction" that will explain the dispersion of light but rather the "theoretical bricks" that, when added together, will provide an explanation of the phenomenon. Schaffner summarizes his position: "The message from this prima facie strong case of intertheoretic reduction is that we get fragmentary and partial explanations of parts of a discipline, but not any type of overall sweeping reduction. The "reductions" are creeping, not sweeping" (Schaffner 2006, 378).

Schaffner's cumulative view of explanation implies that explanation does not come from a process of reduction or deduction from a broad theoretical context but rather from a patchwork of theories that may slowly but surely illuminate phenomena. For philosophers of biology, causal-mechanical explanations often seem much more appropriate than a formal reductive model.[2] It is this observation that allows Brigandt and Love to claim that what is seen as a reduction is very often just an explanation by parts of the so-called reductive theory. A common feature of most of these alternatives to reductionism is an emphasis on the pluralistic dimension of the epistemology of biology. The different parts of the explanation are no longer seen as rivals but as complementary.

Explanatory pluralism, as applied to the life sciences, is thus the idea that any theoretical argument in biology must refer to a variety of methods, models, and modes of action in order to be complete (Mitchell 2003). This view broadly implies that a given type of explanation, which may include several explanantia, corresponds to a single type of phenomenon. But a stronger view of pluralism holds that the same type of phenomenon can be understood not only by different explanantia but also by different kinds of explanations. Most philosophers reject this view, considering that in most cases there is incommensurability between two different kinds of explanation for the same kind of phenomenon. The problem, then, will mainly be to identify the kind of explanation to be ascribed to the type of phenomenon and the way in which the empirical content makes it explanatory (i.e., to identify the different explanantia of the explanatory kind). The same will be true when it comes to explaining variation in biology.

In fact there are different types, levels, and degrees of variation that can be observed in the living world.[3] The very idea of variation is approached differently in different areas of biology. The same type of explanation is not given for variation in populations over time, variation in individuals, variation in a trait for a given species, or genetic variation. The pluralistic approach to the problem of variation seems as appropriate here as elsewhere: the various phenomena listed above all refer to different situations in which the empirical content of the reference differs. Under these conditions it is preferable to refer to different types of explanation depending on the situation. For example, genetic variation is best explained probabilistically, whereas individual variation is best explained causally and mechanistically. In a sense, it seems that the unification or integration of explanations is not an end in itself but necessary to the goal of solving a particular scientific problem, with the nature of the problem determining the type of intellectual integration required (Brigandt 2010, 295). On the basis of Brigand's model, it becomes possible, for example, to explain population variation in terms of genetic variation by means of an explanation that integrates these two sets of phenomena. In other words, types of explanations only emerge in the context of a problem.

In conclusion we can say that the question of explanation in philosophy refers to two competing epistemological positions: that of the unification of knowledge and that of explanatory pluralism. However, I argue that by combining these two approaches, a third epistemological path is possible: that of integration. This idea of integration appears, for example, through a unified but pluralistic explanatory theory such as the synthetic theory of evolution—if the modern synthesis can be considered a theory (see Gayon's critique in 1992 [2019])—but this is beyond the scope of our discussion for the moment. From an epistemological point of view, then, the problem is to identify what contributes to the integration of the different kinds of explanation among themselves. Such integration can be achieved, for example, through smaller epistemic units (concepts, methods, models) that are linked together. I will argue that plastic- terms should play such a role by integrating different kinds of explanations of variation in biology. I will illustrate this with two paradigmatic cases: embryonic induction in development and regulation in physiology.

Embryonic Induction Explains Variation in Developmental Process

WHAT IS EMBRYONIC INDUCTION?

Throughout the twentieth century, scientists have been trying to identify the precise set of factors at the origin of embryonic induction: that particular phenomenon that allows cells of one type to be "transformed" into another type

AN EXPLANANS IN DEVELOPMENTAL PROCESS

and that also allows evolutionary changes to occur through heritable changes during ontogenesis. Although the scale of observation of embryonic induction phenomena has become increasingly refined over time, the observed result has always remained the same: embryonic induction leads to cell differentiation under the effect of a physical or chemical interaction between undifferentiated cells. This ability to differentiate, that is, to generate a different cell state or type in the developing organism, corresponds to what Gavin de Beer called "histogenetic plasticity:" "Another sense in which the term plasticity has been used relates to the ability of a tissue to undergo further or other differentiations. This is histogenetic plasticity" (De Beer 1940, 83).

This concept of "histogenetic plasticity" and its associated definition continue to be linked to another term, *morphogenetic plasticity*, which is somewhat more widely used in the literature. At a symposium of the Society for Experimental Biology in Cambridge in 1948, the geneticist Cyril Darlington explained how a cell's morphogenetic plasticity, or morphogenetic competence,[4] would decrease as differentiation progressed (Society for Experimental Biology 1948, 318). The term *morphogenetic plasticity* was then used to describe the ability to differentiate at both the tissue (*histos*) and cellular levels. Half a century later, the term *morphogenetic plasticity* retains the same meaning, and biologists continue to refer to it, as this extract from a 2005 article in the *Journal of Biological Theory* shows:

> In a natural developmental process cells successively lose their multipotency, that is their ability to produce a set of different cell types. Plasticity is the ability of one cell type to change to another cell type. As the development proceeds, the number of cells and cell types increases, while the plasticity of cells generally decreases. Through the metamorphosis, however, some cells recover the ability to produce other cell types by regaining plasticity. (Takagi and Kaneko 2005, 173)

It is clear that when biologists talk about "multipotency" or "morphogenetic plasticity," they understand "plasticity" as synonymous with a certain "capacity or potency to differentiate" into different types of tissues or cells.

The process of embryonic induction is now commonly defined as "the physical interaction, either directly or through a mediator, between at least two cells with different developmental histories and properties." Other more general definitions can be found in the literature, such as: "Induction is a developmental mechanism in which a cell lineage induces neighboring cell lineages to differentiate (adopt somatic duties)" (Resnik 1992, 456). In any case, induction and morphogenetic plasticity are always linked by the common idea of *cell differentiation over time*.

In light of these different definitions, it is clear that the cells involved in the induction process (the cells emitting the inductive signal, the inductive cells, and the cells exposed to this signal, known as *competent cells*[5]) have differences between them that allow them to be distinguished even though they are all said to be "undifferentiated" at the time of induction. But it is precisely because there are initial differences between these cells—which will communicate at the moment of induction—that an induction process can be initiated. In other words, the initial differences between cells are not caused by the induction process because they precede it. These differences must therefore be related to the history and/or relative spatial position of the undifferentiated cells when the induction process is ready to be initiated. These two aspects—history and/or spatial position—are linked in that the status and topology of a cell change progressively during development. In *Keywords and Concepts in Evolutionary Developmental Biology*, the entry "embryonic induction" is defined as follows:

> Cells involved in embryonic induction are progressing along a series of one-way streets. This process, driven by complex interactions between cells within the embryo, involves intercellular cross-talk and reciprocity that lead to evermore-complicated histological and anatomical structures. Such interactions provide the opportunity to specify the fates of cells based on their position and prior inductive interactions. (Hall and Olson 2006, 103)

This is the developmental process, and unlike the evolutionary process, which, as Darwin pointed out, does not always move toward the most advanced, it is always characterized by progressive specialization, or the accumulation of cumulative variations and differences. The organism at the end of its development is *by definition* more advanced and complex than the original zygote from which it originated. The term often used to refer to this idea of a "capacity to generate difference" is that of *morphogenetic plasticity*. Embryonic induction leads to precisely this result and is seen as a particular causal-mechanical explanation for the origin of variation during development. However, as we will see in this chapter, the induction process is not the only way in which variation between cells is generated during development.

But first, it seems that by studying the process of induction and its specificities, it is possible to consider the extent to which (morphogenetic) plasticity can be understood as an explanans of variation. However, if (as I argue in this chapter) induction as a whole cannot account for this "capacity to generate difference" that characterizes developing organisms, then this implies that plasticity and induction are not exclusively linked. A subsequent question would be, Once the biologist begins to understand the mechanisms of induction more precisely, will she or he be able to avoid any reference to plastic- terms? To answer this question, I will first look at the use of the term *induction* and the criteria associated with it in the biological literature.

Spemann and Mangold were the first to highlight the process of induction in 1923 (see chap. 2). At that time, the molecular determinants of this process were not known. The levels of observation were the organism, the tissue, or the cell. Spemann and Mangold used the term *primary inducer* to characterize an area of the embryo (which was relatively imprecise at the time) that was thought to be the cause of the crucial modification of embryonic cells during development—their "determination" into specific cell types. Since the primary inducer is defined by the area where cell determination is thought to occur, it would become the focus of attention for embryologists interested in identifying and understanding the fine details of the underlying mechanisms.

The molecular factors at the origin of the induction process were not identified until the middle of the twentieth century. However, the transition from the organism (or cell) to the molecular level did not mean the abandonment of the concept of induction, which has remained in the vocabulary and textbooks of biology to this day. In the sixth edition of the textbook *Developmental Biology*, Gilbert refers several times to the term *induction* and explain its general action the following way: "The induction of numerous organs is affected by a relatively small set of paracrine factors[6] that often function as morphogens" (Gilbert 2000a). The term *morphogens* is from the Greek and means "form-givers." They are diffusible biochemical molecules that can determine the fate of a cell by their concentrations, in that cells exposed to high levels of a morphogen will activate different genes (throughout signals) than those exposed to lower levels" (Gilbert and Barresi 2016, G-20).[7]

One of the probable reasons why the concept of induction has persisted in the vocabulary of biologists is that it allows a number of competing molecular models—but all based on the same types of effects at the cellular level—to be brought together and integrated under a single generic name: *embryonic induction*. The generic name is used to refer to the singular effect it produces at the cellular level (differentiation process). This observation is at most a pragmatic justification for the presence of the term in the literature. It is possible to explain a variety of molecular mechanisms by referring to a single term. However, if it is possible to pragmatically justify the persistence of the notion of induction, I can venture to propose a more theoretical explanation of the validity of this notion by identifying specific criteria for induction.

To do this, it is first necessary to establish a classification to target the induction criteria. A distinction can be made between molecular mechanisms of induction (MMI) (specific to the induction process) and molecular mechanisms before induction (MMBI) (more general to the development process).

It should be noted that MMBI can be divided into two subtypes: MMBID (generic mechanisms that generate cellular *difference* or variation) and MMBIS (generic mechanisms that do not generate cellular difference or variation—S for *similarity*). An example of MMBIS is provided by molecular mechanisms associated with mitosis that do not produce cellular difference or variation, since mitosis by definition produces two daughter cells that are similar to the parent cell. Conversely, an example of MMBID is the *differential* expression on the cell surface of proteins responsible for cell adhesion, an expression that can generate significant cellular differences. For example, the polarity of certain cells is known to be due to this differential expression on the cell surface of proteins responsible for cell adhesion: the resulting cells from the division of such a polarized cell could be an adhesive cell and a nonadhesive cell.

The two examples of MMBIS and MMBID, it should be emphasized, are not, strictly speaking, induction phenomena (MMI) but phenomena that precede induction in time and can be the source of differences (MMBI). Therefore, in order to understand the specificity of MMI, it is necessary to highlight what distinguishes MMBID from MMI.

Induction Explains the Ability to Create a Difference in the State of a Cell by Means of Intercellular Signaling

The main characteristic of MMI, in contrast to MMBID, is that it generates new cell states by means of intercellular signaling. In *Biological Physics of the Developing Embryo*, Gabor Forgacs[8] and Stuart Newman suggest that there are two types of mechanisms by which new cell states can be generated:

There are two major categories of basic pattern-forming mechanisms that produce patterns by cell differentiation, that is, by generating new cell states: those that do not depend on cell-cell signaling (cell "autonomous" mechanisms) and those that do ("inductive" mechanisms) (Forgacs and Newman 2005, 159).

What Forgacs and Newman call cell autonomous mechanisms corresponds to MMBID in our terminology. For example, a new cell state may result from the division of a heterogeneous egg cell. It is considered to be heterogeneous because it contains molecules that are differentially distributed in the lumen of the cytoplasm—these are morphogenic gradients. After cleavage, the cell produces different blastomeres[9] as a result of this initial heterogeneity (fig. 6.3).

The creation of a new cell state can also be due to asymmetric mitosis. In this case, different molecules have been transported by intracellular processes into the two halves of the parent cell, leading to the generation of two different daughter cells when the cell divides by mitosis (fig. 6.4).

These two examples suggest that, in the case of MMBID, it is molecular and physical processes that do not involve intercellular signaling that produce cellular differences over time. However, if I transpose these two mechanisms

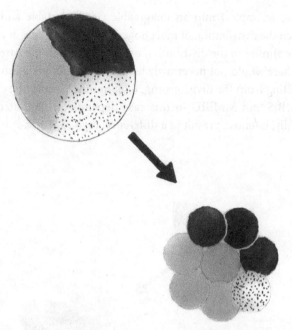

FIGURE 6.3. Mechanism generating a new cell state without induction signaling: cell "autonomous" (or MMBID). In this case, the state of the new cells are simply the result of the divisions of the initial heterogeneous egg cell. The heterogeneity of the daughter cells is due to the morphogenic gradients of the initial cell. Drawings by Anton© based on Forgacs and Newman (2005, fig. 7.1a, b), itself based on Salazar-Ciudad, Jernvall, and Newman (2003, fig. 1).

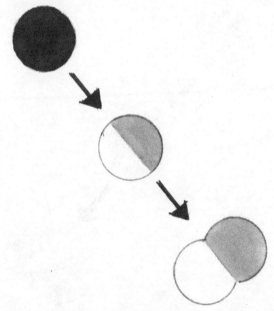

FIGURE 6.4. Drawing illustrating a mechanism creating new cell states without signaling: "cell autonomous" (or MMBID). The new cell states are the result of asymmetric mitosis, leading to a heterogeneous molecular distribution between the two cells. Drawings by Anton© based on Forgacs and Newman (2005, fig. 7.1a, b), itself based on Salazar-Ciudad, Jernvall, and Newman (2003, fig. 1).

(the same two examples) into an imaginable (ideally) stable and ordered world in which the distribution of morphogens in daughter cells at each division would be similar to the distribution of morphogens in the parent cell, it is clear that there would not necessarily be a difference or variation between the cells resulting from the divisions (fig. 6.5). In other words, the difference between MMBIS and MMBID in this case depends on the outcome. But probabilistically, it tends to result in a difference, ergo MMBID.

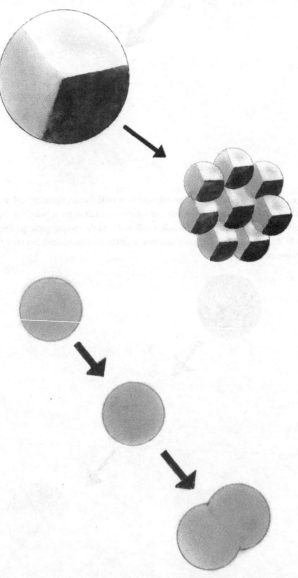

FIGURE 6.5. Drawings illustrating an "ideal" situation in which mechanisms (of the "cell autonomous" type) that could create a new cell state without signaling do not do so. (*Top*) Distribution of morphogens in daughter cells identical to the distribution of morphogens in the mother cell after division. (*Bottom*) Mitosis leading to a homogeneous molecular distribution between the two daughter cells. Drawings by Anton©.

To illustrate this "ideal" case with a concrete situation, let us take an example. The biologist Hans Meinhard took one of the rare exceptions of an almost homogeneous egg: the brown seaweed *Fucus*. Unlike the eggs of most species, which mature in an asymmetric environment and quickly become inhomogeneous, the development of the spherical egg of the brown algae only begins with the outgrowth of a pseudorhizoid from the egg (fig. 6.6). The appearance of this outgrowth can be very late because of the homogeneity of the egg. It is only the appearance of a pseudorhizoid that allows development to begin. This example illustrates that development requires heterogeneity before (and even for) intercellular signaling to occur. In the case of brown algae, the outgrowth can then be directed in one direction or another depending on differences in temperature, pH, light, or an electrical field applied to it (ionic differences). Somehow a localized change will trigger development.

This example also illustrates the fact that this homogeneous egg is in a rather unstable situation, and in this context any asymmetry could guide the formation of the complex structure. In the absence of any orientation effect, the outgrowth will appear randomly but with a large delay. We could argue

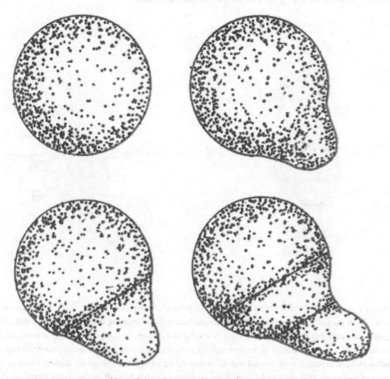

FIGURE 6.6. Figure proposed by Jaffe (1968), repeated by Meinhard (1982, 25–26), and copied by Anton© to illustrate the development of the initial homogeneous spherical egg of the brown seaweed *Fucus*.

that the probability of homogeneous division is much higher there than in other organisms, that it takes more events to have a heterogeneous cell.

In contrast to this set of mechanisms, Forgacs and Newman (2005) and Salazar-Ciudad, Jernvall, and Newman (2003) refer to "inductive mechanisms" (what I have called MMIs) and distinguish situations where the creation of a new cell state depends on *cell signaling*. For example, a cell produces a signal to a neighboring cell, which changes its cell state in response to this signal (fig. 6.7).

In the case of MMI, therefore, the situation is different from that of MMBID because the induction mechanism *necessarily* leads to variation between cells at the end of the mechanism, whatever the ideal world considered or that can be invoked. For there to be no variation between cells, cell signaling—the interaction between cells—should be "silent." In this case, both cells (the inducing cell and the competent cell) would remain as they were before they came into contact with each other. In other words, with MMI, there must be cellular signaling to obtain an effect—whatever it may be—whereas with MMBID, there is no need for differential distribution of molecules to obtain a result (see fig. 6.7). Without differential distribution, at least two cloned daughter cells could be obtained.

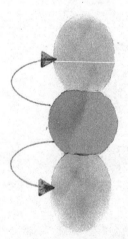 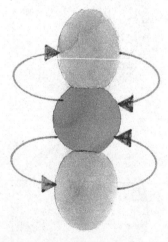

FIGURE 6.7. Drawing illustrating mechanisms creating a new cell state with signaling: "induction mechanisms." On the left is a "hierarchical induction mechanism": the inducing cell (*center of the stack*) affects neighboring cells, but the induced cells (*cells on the top and bottom*) do not produce a signal in return. On the right is an "emergent induction mechanism": the inducing cell (*center*) affects neighboring cells (*top and bottom*) which, in return, emit a signal that influences the initial signal. Drawings by Anton© based on Forgacs and Newman (2005, fig. 7.1a, b), itself based on Salazar-Ciudad, Jernvall, and Newman (2003, fig. 1).

The nuance between MMBID and MMI is fine but important. There seems to be an ontological difference between MMI and MMBID that requires an epistemological distinction to be made. In the first case, MMBID can *possibly* (fig. 6.5) make a difference (and it will in the vast majority of situations; see figs. 6.3, 6.4). The new cell state can therefore be described as *probabilistic*. In the second case the MMI must *necessarily* create a difference (fig. 6.5). The new induced cell state can therefore be said to be *necessary*. This distinction allows us to conclude that the induction mechanism is a phenomenon that allows the creation of actual differences (in the sense of actual opposed to potential or probabilistic). In this sense it is a specific explanation for the creation of variation in development. The process of induction cannot be considered as a particular statement, one among others, in the explanans of variation in the developmental process because it constitutes a *necessary* explanation for the generation of difference in the developmental process. However, one might ask whether there are other mechanisms besides MMI that help to explain the generation of differences in the developmental process (just like MMBID) but that (unlike MMBID) would have other criteria in common with induction, such as interaction with neighboring cells. More specifically, I will consider what distinguishes plasticity from induction. The first distinction that needs to be made between plasticity and induction is that plasticity is a disposition (i.e., it can designate a cell as capable of participating as a respondent in the process) and induction is the process. Once this first consideration has been highlighted, it remains to understand what is specific to the induction process.

Induction, an "Activation" of the Cell?

The process of embryonic induction is often described as an active process in the sense that it is identified not only by its cause—the fact that it brings two cells together—or its effect—the ability to create a new cell state—but also by describing what links cause and effect: the nature of the mechanism involved. This is why the term *induction* is sometimes used synonymously with *activation* (Forgacs and Newman 2005, 169). This observation helps us understand the assimilation that biologists often make between the processes of induction and those of activation. However, if in the field of cell physiology, the term *activation* obviously refers to an interaction between cells, can it be said that there is a creation of variation in shape (pattern formation) when there is a cell activation? In other words, are all induction processes also activations? Or is "activation process" good enough to describe induction?

The term *activation* is often associated with a degree of specificity, for example, when biologists speak of "molecular activation" or "cell activation."

The problem is that in the case of activation, the degree of specificity is usually given, whereas in the case of induction, the degree of specificity is rarely given (the level of observation is usually the cellular level). Does this mean that cell activation is identical to induction, or is it rather molecular activation?

At the molecular level, activation can be seen as one possible type of signaling among several mechanisms responsible for cell induction. However, this type of signaling (molecular activation) alone does not account for all possible molecular signaling mechanisms responsible for induction. For example, inhibition is the other possible molecular signaling mechanism for induction.

At the cellular level, there are also differences between activation and induction. As we have seen, induction creates/generates a new cellular state. This is not systematically the case with cellular activation (I will show examples of this later). So why are "activation" and "induction" often confused when their underlying processes are different? The main reason is that at the molecular level, both inhibition and activation are observed as phenomena responsible for cell activation, although activation is not necessarily attributed to them on a smaller scale.

Since the process of induction does indeed involve intercellular signaling (see chap. 5), the assimilation between induction and activation is in fact based on the tacit confusion between the idea of intercellular signaling and that of cell activation, on the understanding that the cell is considered to be activated because it is stimulated by a molecular signal. In immunology, for example, the term *activation* is used to describe the signaling that takes place between immune cells and that allows, in particular, the initiation of the immune response by T cells. If the induction phenomenon is accompanied by cellular stimulation, then the competent cell "senses" the inducing signal and responds to it. However, this does not mean that molecular activation as a possible mechanism of induction is a necessary mechanism of induction.

Why is this an essential point for understanding the theoretical importance of the concept of induction? By confusing signaling or molecular activation with cellular activation and by equating these different mechanisms with the induction process, there is a danger that any "molecular activation" or "cellular activation" identified by the biologist will be seen as an expression of an induction process. In fact these are ontologically different processes that will produce different biological effects. This false assimilation leads some contemporary biologists and theorists of biology to believe that any molecular activation must lead to a change in cellular state and, by extension, that any molecular activation generates variation that is determinant for development. This ontological confusion leads to an epistemological error, since the process of induction is no longer simply a mechanism specific to morphogenesis or histogenesis—as its original use, with Spemann's embryonic induction, seemed to indicate—but refers to all "activation" mechanisms during the life of the organism because activation mechanisms—both at the molecular and cellular level—occur at any moment in the life of an organism and are not limited to its morphogenetic or embryonic development.

Many common physiological phenomena involve activation mechanisms that involve chemical processes. This is particularly true in plant physiology, where many chemicals can cause molecular activation. For example, when the stomata of the leaf cells of a water-stressed plant close, this is seen as a change in its physiological state (fig. 6.8).

In the example described above, activation, sometimes called inducible plant defense mechanism, does not involve any change in the morphological nature of the cell: the stomatal cell remains a stomatal cell, only its physiological state is modified (the modifications are within normal parameters,

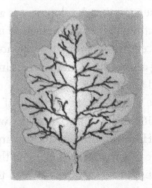

FIGURE 6.8. Drawings depicting the changes that can occur in stomata opening (*top*) and vascular activity (*bottom*) of a leaf as a function of water availability in a plant. Drawings by Anton©.

functionally identical, so it is not a variation in the sense that a novelty is generated), and there is no intercellular signaling as in embryonic induction. This is indeed an activation mechanism, but it is a far cry from the original definition of embryonic induction as a process that creates a new cell state or a process that causes developmental variation.

Another example where induction is equated with activation is "differential cell adhesion" (Forgacs and Newman 2005), which also initially depends on molecular activation. In this case, as in induction, there is an interaction between neighboring cells. The cells send signals and communicate with each other through direct contact. However, this interaction does not lead to a change in the overall morphological state of the cells involved, but only to a change in their organization relative to each other.

These examples show that there is an ontological difference between activation—whether at the cellular or molecular level—and induction, since activation does not explain the creation of developmental variation in the same way as embryonic induction, and in particular it does not explain the ability to create a difference in cellular state. Therefore, the distinction between the notions of induction and activation seems justified for certain situations.

If the notion of induction cannot simply be equated with that of activation, can we say that there are other, more general criteria that allow the emergence of developmental variation to be explained in the same way as induction? To answer this question, I will focus on a central criterion of the induction process: the temporal hierarchy that exists within the process.

Time Hierarchy Determines Induction

In Brian Hall and Wendy Olson's definition of "embryonic induction" (quoted above) in *Keywords and Concepts in Evolutionary Developmental Biology*, the authors suggest that development is a gradual and hierarchical process. They argue that the inductive interaction is itself hierarchical. Thus, there are two kinds of cells in any inductive interaction—responding cells and signaling or inducing cells—and at many points in the process their roles may be reversed (Hall and Olson 2006, 103–8).

As we have seen, the process of induction generates physical and molecular differences or variations between the cells involved in the induction process. However, the authors also argue that another type of developmental difference or variation is generated that is related to the differential relationship established over time between the inducing cells on the one hand and the responding[10] or competent cells on the other. This differential relationship depends on the temporal hierarchy in which the different roles of cells during the inductive interaction are articulated. Inductive cells play the primary role during the induction process. As their name suggests, responsive cells act only in response to the signal sent by inductive cells. These specific roles, assigned at a given time for specific cells within a given inductive relationship, are therefore not fixed once and for all. Roles are contextual. The hierarchy depends on the specific temporality of development, and the respective status of the cells is therefore chronohistorical. This hierarchical aspect within the induction process is a criterion that has often been discussed among biologists and theorists interested in the induction process.

There is a hierarchy within the process of induction, but there is also a hierarchy between the different processes of induction within development in general. This second hierarchy is manifested by the existence of so-called critical stages of development (e.g., gastrulation[11] or neurulation[12] in triploblastic organisms[13]), without which full development, or simply the transition from one stage to another, is not possible (see fig. 6.9).

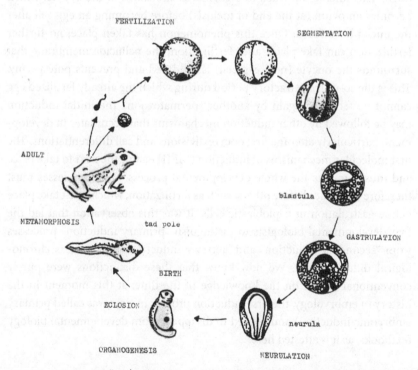

FIGURE 6.9. Drawing illustrating the life cycle of a triploblastic organism, the tetrapod frog. After fertilization, the egg undergoes cell division before passing through a developmental stage of gastrulation (which allows it to develop into three embryonic layers), then a stage of neurulation (establishment of the primitive nervous system) before the onset of organogenesis. All of these stages are considered crucial developmental steps, without which the generation of a viable individual cannot take place. Drawing by Anton©.

If one of these steps is not done correctly, development stops. In a sense, the temporal hierarchy of the induction process explains the gradual variation that occurs during development. As Roger Sawer and Loren Knapp point out in their definition of embryonic induction:

> Fertilization may be viewed as the first hierarchical level of induction [in metazoans]. The specialized sperm cell activates the egg and sets in motion the differential gene expression that will transform the zygote into an adult organism (Sawyer and Knapp 2006, 104–5).

In the first step necessary for the induction process—fertilization—we find the criteria described above (i.e., the interaction between two cells with an exchange of signals and the creation of a new cellular state): two "different" cells (the oocyte and the spermatozoon) enter into a relationship at the end of which their nuclei fuse (so there is a signaling or interaction between the cells). This interaction leads to a change in the status of the oocyte, which becomes an ovum (at the end of meiosis) before becoming an egg cell after the nuclei have fused. Once this phenomenon has taken place, no further fertilization can take place: after fertilization, the pellucida membrane that surrounds the oocyte (now egg cell) is modified and prevents polyspermy. This is the so-called refractory period during which the already fertilized egg cannot be fertilized again by another spermatozoon. The initial induction may be followed by other induction mechanisms that occur later in development, particularly after the first oocyte divisions and cell differentiations. The first molecular mechanisms of induction (MMI) enable the next to take place and thus complete the whole developmental process. Some processes must therefore take place before others, such as fertilization, which must take place before gastrulation in triploblastic cells. It was this observation that led the first developmental biologists to distinguish "primary induction" processes from "secondary induction" and "tertiary induction" processes as chronological differentiation. We now know that these distinctions were purely conventional (based on the knowledge of the time: at this moment in the history of embryology, the first induction process in time was called primary embryonic induction) and tended to disappear from developmental biology textbooks, as it is attested here:

> Because there are numerous inductions during embryonic development, this key induction—in which the progeny of dorsal lip cells induces the dorsal axis and the neural tube—is traditionally called the primary embryonic induction. This classical term has been a source of confusion, however, because

the induction of the neural tube by the notochord is no longer considered the first inductive process in the embryo. (Gilbert and Barresi 2016, 347–48)

Although the conventional distinctions between primary, secondary, and tertiary induction are tending to disappear (although their chronological order remains a reality), there is still a temporal hierarchy in the inductive processes reflected in the expression "cascade of inductions," which in turn is linked to a cascade of gene expression. In fact, the expression of genes can initially affect the whole organism and then be spatially restricted to certain regions (). This hierarchical gradation in the time of induction helps to explain the progressive loss of *developmental potential* as development proceeds.

The notion of *developmental potential* is often expressed in the literature on stem cells in relation to the notion of plasticity (e.g., Rinkevich and Matranga 2009; Ettenshon et al. 2007; Galliot et al. 2006, etc.). The idea is that the earlier the stem cell is collected in the life of the organism, the more "plastic" it is and the higher its differentiation capacity remains. Traditionally, a distinction has been made between "totipotent" stem cells, which characterize the egg cell (the zygote) up to the eight-cell blastomere and can give rise to all cell types (i.e., an entire organism); pluripotent stem cells, which can give rise to all cell types except the embryonic appendages; multipotent stem cells, which can give rise to different cell types but are specific to a particular cell lineage; and unipotent stem cells, which can give rise to only one cell type. There are still a number of controversies regarding the exact developmental state and capacity of each of these cell types (see Laplane 2016). Nevertheless, general differences are observed, clearly demonstrating the existence of a hierarchy of "developmental potential" that in turn correlates with the temporal hierarchy of cell appearance.

What is the biologist's interest in using the term *plasticity* rather than the process of induction? When they speak of morphogenetic plasticity, they emphasize the ability of a cell to induce a differentiation mechanism, or the ability of a cell to respond to this induction (e.g., Furusawa and Kaneko 2006). On the other hand, the term *morphogenetic plasticity* does not allow them to emphasize the hierarchical relationship between the two cells. The important distinction, then, is between induction as a hierarchical process in time and plasticity as a dispositional property of a given entity (or disposition to induction). The phenomenon of induction generally refers to the process as a whole (to a signaling relationship), whereas plasticity generally refers to the property of a given entity: a tissue or a cell. In this sense, the concept of induction seems to provide a more specific explanation of the cause of variation in development than the concept of plasticity.

From Induction to Plasticity: How Plasticity
Helps Explain Variation

For the biologist, the reference to plasticity implies being part of a theoretical explanatory scheme in which she will try to identify and explain the variation in the developmental process, in the broad sense, from a set of diverse phenomena. The reference to the induction process provides a specific empirical content that will be necessary for an adequate explanation (that respects Hempel's DN model) of developmental variation in biology. It is a kind of method by which plasticity is replaced by "determination." In the case of the induction process, the proposed explanation implies a low level of generality but retains sufficiently broad criteria to be applied to different types of mechanisms observable at the molecular level.

Therefore, the explanation of developmental variation by induction depends on the identification of several criteria that we have discussed above, such as the interaction (or signaling) between cells or tissues of the developing organism. The specific induction explanation will therefore apply to a limited number of phenomena. For example, developmental variation phenomena resulting from the activation of a single cell will not be considered to be explained by the process of induction.

Biological Plasticity, a Synonym for Biological Regulation?

The concept of regulation involves at least three ideas:

(1) that of the interacting relationship between unstable elements;
(2) that of the criterion or benchmark;
(3) that of the comparator [*comparateur*].

Regulation is the adjustment, in relation to a few rules or norms, of a plurality of move-
ments or acts and of their effects or products, which their diversity or succession ren-
ders initially alien.

GEORGES CANGUILHEM (1968, MY TRANSLATION)

Regulation Explains Variation in Physiology

The concept of "regulation" has been used in science since the beginning of
the twentieth century, but its use in biology has only increased in recent years,
particularly with the observation and description of mechanisms that control and correct genomic information. It is also the identification of these
so-called regulatory mechanisms that has led to the gradual abandonment
of a linear model linking genotype to phenotype in favor of network models linking a set of genetic and epigenetic factors (intranuclear, such as DNA
methylation or histone modification) to a given phenotype. These discoveries
have gradually shifted the debate on regulation from the cellular and organizational level to the molecular level.[1]

In embryology the term *regulation* was already widely used in the early
1900s. The embryologist Charles Manning Child (1869–1964), for example,
wrote a series of "Studies on Regulation" between 1901 and 1905. The term
regulation was then used by Child mainly in the sense of "regeneration." In
an article published in 1906, in an attempt to defend a "theory of regulation,"
Child states that regulation can be defined as "the return or approach to physiological or functional equilibrium after such equilibrium has been disturbed"
(Child 1906, 424). This relatively broad definition is, however, clarified by
Child's specification of two different types of regulation: "redifferentiation"
and "regeneration," which he distinguishes on a functional basis (simple cellular differentiation in the first case or restoration of an impaired function in

the second). It is the spatial conditions of the organism and its environment that induce one or the other process.

With this embryological connotation, regulation seems to be closely related to embryonic induction but also to developmental plasticity as the ability to generate variation in form. Does regulation explain the generation of variation as well as induction? I will first address this question before considering the relationship between regulation and plasticity.

<h2>WHAT IS REGULATION?</h2>

At the beginning of the twentieth century, the term *regulation* was used in embryology mainly to characterize the ability of an early embryo to continue and restore its development after its structure has been altered or damaged. Today, this meaning of the term has been largely abandoned in favor of the more precise terms *regeneration*[2] and *dedifferentiation*[3] (or *redifferentiation*). But the word *regulation*, because of its polysemy (it can be understood as homeostasis, for example, or as in "to make regular"), has also been used in the last century in a less specific context than that of embryology. Indeed, the polysemy of the word explains why its use in biology has varied according to the context. For example, one of the earliest definitions of the term in biology was relatively close to contemporary definitions of regulation: as an adaptive response of the structure and form of the organism to the conditions it encounters. This idea of "adaptive response" suggested that regulation was the source of variation from an initial state.

It is a historical fact that long before the notion of regulation emerged, the notion of plasticity was already commonly associated in embryology with the idea of regenerative capacity. In fact, at the beginning of the twentieth century, in Child's studies on regulation, the adjective *plastic* often appears as a synonym for *regulatory capacity* (of regeneration). Even today, the term *plastic* is still used in a biological context where someone is evaluating the regenerative capacity of certain tissues, in particular stem cells. The idea that stem cells have the capacity to retain a certain plasticity, that they are the locus of remanent plasticity, is found in contemporary literature. Therefore, the link that still exists today between the concepts of regulation and plasticity can be seen as a survival, at least in part, of the earlier link that existed in embryology between the concepts of regeneration and regulation. Nevertheless, the question is whether all references to regulation at the cellular level in the biological literature can be expressed in terms of plasticity and whether they refer exclusively to this regenerative mechanism (i.e., the ability to generate a new form), or whether, on the contrary, they cover other types of mechanisms. In other words, is regulation, at the cellular level, always plasticity? Is it always regenerative? Can it not also be degenerative?

Regulation, from Development to Physiology

If in embryology the notion of regulation is often linked to that of regeneration, it is because morphogenesis and development have long been the two privileged objects of study in the life sciences. Moreover, embryology is one of the first fields of biology to use the concept of regulation. As Georges Canguilhem (1904–1995) has shown in *Idéologie et rationalité dans l'histoire des sciences de la vie* (*Ideology and Rationality in the History of the Life Sciences*), it was not until the last third of the nineteenth century that biological functions that control other biological functions were given the name of "regulation." In physiology, the term *regulation* was first introduced only through metaphors that referred mainly to physical phenomena (Canguilhem 1981).

For example, Claude Bernard (1813–1878), considered the father of modern physiology, did not use the term *regulation* until 1878, and then only very sporadically, to describe the "regulator," a concept derived from physics that describes a function of conservation and restitution in closed systems (Bernard 1876). Regulation in biology allows Bernard to think about the way in which deviations from the norm that occur in the *milieu intérieur* (intracellular environment) can be compensated. This view of regulation introduced a new context, different from embryology, into the field of life sciences: that of biophysics. But Bernard's original concept of regulation also differs from other views of regulation proposed at the same time, such as that of Auguste Comte (1798–1857). According to Comte, regulation is a process by which the outside regulates the inside (regulation somehow maintains order) (Comte 1876). For example, the stability of the solar system allows living systems to be regulated though the mediation of the environment.

The differences between Bernard and Comte in their vision of regulation illustrate two aspects that are still common to the concept of regulation in biology. Regulation is either seen as an intrinsic property of living systems (e.g., the property of homeostasis, which implies a return to equilibrium within the system) or as an emergent property of living systems (due to their interactions with the external environment in which they develop and evolve).

This possible duality of the term *regulation* is reminiscent of the duality of the term *plasticity* between its active and its passive meaning. If, since the middle of the nineteenth century, the definition of the term *regulation* no longer seems to be limited to the sole domain of embryology (and regeneration), the question remains: is the notion of plasticity, which is also found in the physiological literature, also linked to these new definitions of regulation, understood either as a "specific capacity of living organisms for self-regulation" (a potentiality) or as an "observable result of the interaction of living organisms with their environment" (an output). In the next section, I first consider what happens when regulation is understood as an observable outcome of the interaction of living organisms with their environment before examining the situation when regulation is understood as a potentiality. The examination of these two understandings (one after the other) will then allow me to address the main question at stake here, namely, whether regulation can explain the generation of variation in the same way as induction by highlighting the possible links between plasticity and regulation.

Regulation Determined by the Interaction of Living Systems with Their Environment

The question of regulation in the context of the interaction of living systems with their environment leads us to examine the work of the biologist, psychologist, and epistemologist Jean Piaget (1896–1980), for whom this question occupied a very important place in his thinking. Since the 1910s the relationships and interactions within a complex system have been a central question for many theorists and philosophers. The structuralist school of thought, inspired by a linguistic model initiated by Ferdinand de Saussure (1857–1913),[4] tends to understand the whole of social reality as a formal set of relations. The structuralist school of thought will be one of the sources of systemics, a science that emerged in the 1950s and that examines its objects of study according to a global (or holistic) approach. Systemics is the oldest of the direct branches of structuralism, among which we can mention cybernetics or general systems theory (to which I will return later and for which the question of regulation is central).

In 1968, almost fifty years after the beginnings of structuralism, Piaget showed that, contrary to what some of its theorists had claimed, it should be considered a method and not a doctrine: "The permanent danger that threatens structuralism . . . is the realism of the structure that we end up with as soon as one forgets its connections with the operations from which it is derived" (Piaget 1968, my translation here and elsewhere for Piaget). But he takes a constructivist view of structure, saying that "there is no structure without a construction, abstract or genetic":

As a first approximation, a structure is a system of transformations, which includes laws as a system (as opposed to the properties of elements) and which is preserved or enriched by the very play of its transformations, without these transformations ending outside its boundaries or calling upon external elements. In a word, a structure encompasses the three characters of totality, transformations, and self-regulation [*autoréglage*]. (Piaget 1968, 6–7)

For Piaget, the definition of a structure depends on three fundamental characteristics: totality, transformations and self-regulation (and it is, of course, this last notion that will be of particular interest to us later). As far as the idea of totality is concerned, it implies that if a structure is made up of elements, these elements are subordinated to laws that characterize the system, and "these so-called laws of composition are not reduced to cumulative associations but confer to the whole itself overall properties distinct from those of the elements" (Piaget 1968, 8–10). The second characteristic, the system of transformations, is the characteristic of any structuring activity that, according to Piaget, should not be considered, because of its stability, as an entity (in biology or in the social sciences) or as something innate (in psychology) but that can be explained thanks to the constraining processes of equilibrium. Finally, the third fundamental characteristic of structures is that they are self-regulating. For Piaget, self-regulation leads to the maintenance of structures and to a certain closure. He writes, "Starting from these two results, they mean that the transformations inherent in a structure do not lead outside its boundaries but only produce elements that still belong to the structure and retain its laws" (Piaget 1968, 13). According to Piaget, this is what defines "perfect" regulation, which not only corrects errors afterward but manages to "precorrect" them, so to speak, thanks to internal means of control (what is now commonly called a regulatory loop or feedback). In Piaget's view, nothing ever seems to be completely outside the system. The structure is always embedded in a larger structure that indirectly enables its construction through its multiple regulations. This is precisely why Piaget prefers the term self-regulation (*autoréglage*) to regulation.

In 1968, the same year that Piaget wrote about structuralism, Georges Canguilhem proposed for the *Encyclopaedia universalis* a very different definition of the concept of regulation, one that is (partially) removed from the structuralist context or that at least does not refer to it explicitly (see chapter epigraph). For Canguilhem regulation is thus based on an initial heterogeneity that is compensated after a connection, an interaction, has taken place. In this respect Canguilhem stresses that regulation has become "a concept of biology, after having been only a concept of mechanics, [and] while waiting to become a concept of cybernetics, through the mediation of the concept of homeostasis" (Canguilhem 1981, 99, my translation).

A few years earlier, in his book *Introduction à l'épistémologie génétique* (Introduction to genetic epistemology), published in 1950, Piaget had proposed the concept of "equilibrium" to take account of all the possible forms of "adjustment" that Canguilhem speaks of (in the quoted text above) and in particular for two fundamental forms in Piaget's work: "accommodation" and "assimilation." Piaget considered these adjustments to be adaptations in the sense of "process adaptation" and not in the sense of outcome, that is, as "a passage from a less stable equilibrium to a more stable equilibrium between the organism and the milieu" (Piaget 1950, 81), in short, a kind of regulation according to Canguilhem's definition. Piaget defines *accommodation* as a modification of the organism that allows it to adapt to external conditions. The process of accommodation serves to enrich or extend the organism's scheme of action by making it more flexible. Accommodation mechanisms are determined by the organism and have the effect of modifying behavior in order to adapt it to the requirements of the situation. Thus, the external environment determines accommodation.

It is interesting to note that this notion of accommodation has been used in recent years by theoretical evolutionary biologists (such as Mary Jane West-Eberhard, an entomologist by training) to account for mechanisms similar to those described by Piaget (himself mainly about intelligence in psychology) but at the level of all the phenotypes (characters) observable in nature (West-Eberhard 2005). The term *phenotypic accommodation* used here refers directly to a suggestion made by Piaget himself. West-Eberhard defines it as the "adaptive mutual adjustment among variable parts during development without genetic change" (West-Eberhard 2003, 51). According to West-Eberhard, this property of accommodation confers a certain "developmental plasticity" on living systems, such as organisms, by reducing the negative effects of external stresses (constraints) on them. She emphasizes what she calls the "developmental plasticity of the phenotype" (see chap. 8). From this perspective, organisms are seen as capable of changing their phenotype in response to the environment.

This is the case, for example, with some species that are capable of successively changing their biological status from male to female. For example, the clown fish, a so-called protandrous organism, can change from male to female biological status. The largest clownfish in the colony, the one that socially dominates the group and lives near a sea anemone (recall that the clown fish has a mutualistic relationship with the sea anemone), is usually a female. It has functional ovaries and degenerated testicular tissue. There is usually only one female in the group and the males are all smaller. Unlike the females, the males have functional testes but latent ovaries. The dominant male of the group is the only one who is sexually active; the others are subadult males (remaining in the preadolescent state) who are inactive in the reproductive process. However, in the unfortunate event that the female disappears, the dominant male's gonads cease to function as testes and his ovaries become active. He then becomes female. The role of the reproductive male is then taken over by the largest of the subadult males. The transformation is rapid: the new female can start laying eggs as early as the twenty-sixth day after her sex change.

This very atypical example of phenotypic plasticity opens up the possibility of a more general model. It is in this perspective that West-Eberhard takes up the idea of phenotypic accommodation (from Piaget, among others, but it should be noted that the notion had already existed among naturalists since the end of the nineteenth century, where it was relatively close to the notion of acclimatization, i.e., a modification of a form in a particular environment) in order to generalize it and transform it into a theory of variation at the origin of evolution. West-Eberhard's idea is that the forms resulting from a modification of the "phenotypic accommodation" type can then be subjected to natural selection and finally be definitively fixed by another mechanism: "genetic assimilation"—a phenomenon also inspired by Piaget's prolific inventive audacity although developed in a purely psychological context to which I now return.

The second type of adaptation (in the sense of regulation or adaptation process), which Piaget called assimilation, is defined as the integration by the organism or the subject of environmental elements that are initially external to it. This notion of assimilation was first used in biology to denote a certain capacity, possessed by the organism, to absorb or integrate elements that are initially foreign and external to it (a classic example is that of nutrition, where the body "assimilates" food). This notion of assimilation is old in physiology, but Piaget uses it in psychology to designate the capacity of children to internalize or naturalize behaviors that are initially foreign to them. According to him, the ability to assimilate depends on an innate component; what can be assimilated depends on preexisting structures. (There is a kind of assimilability or disposition to take on new ways of being, but not just any way. It is not a general kind of flexibility but a disposition to certain kinds of change.) In any case, for Piaget assimilation is the motor of adaptation processes (adaptation process or regulation).

Within the biological disciplines, however, the concept of assimilation is not confined to physiology; it has also been made famous in genetics, for example, by Waddington's use of the term *genetic assimilation* (see chaps. 3, 7) to describe how a phenotypic trait, acquired as a result of a developmental process and whose variation is induced by the environment, then acquires a genetic basis, which may then be selected in a population and eventually become a fixed characteristic of that population depending on the adaptive capacity of the varied phenotypic trait in the population (Waddington 1953). This phenomenon of genetic assimilation itself appears to be based on a number of rather complex biological processes and to be part of the mechanisms responsible for evolution. Waddington's use of the term *assimilation* seems more metaphorical than Piaget's in physiology: genetic material, unlike food, does not come from outside the organism. However, some similarities can be found between the use in psychology (the act of acquiring cognitive abilities) and in genetics (the act of acquiring a genetic basis for a trait in a population) even though in the first case the use is limited to a subject, whereas in the second case it applies to a trait in general and to a population. Some of these similarities emerge when we look at Scott Gilbert's definition of *genetic assimilation* (Gilbert 2000b, 731–33). Let's take a closer look at his definition.

According to Gilbert, genetic assimilation occurs when four components come together.

First, "The genome must be responsive to environmental inducers." In other words, there must be an *innate* (to use Piaget's term) competence, a disposition of the genome to integrate and transform information from the environment.

Second, "The competence to be induced must be transferred from an external inducer to an internal, embryonic inducer." This is a component whose explanation requires reference to a causal relationship between levels, as one moves from one level of organization to another (from the organism as a whole to an intraorganismal molecular scale). This is the most problematic point for the biologist, since any reductionist approach is generally unsatisfactory. Some answers to this question have been provided in recent years with the identification of regulatory genes, or RNAs, that control the expression of several other genes (and therefore sometimes entire phenotypes) and whose action can be modulated by environmental factors during development. This component clearly corresponds to what Piaget called the self-regulation (*autoréglage*) of the system. This is the second level of regulation, involved in the construction of new structures that encompass the previous ones and integrating them as a substructures within larger structures.

Third, "There has to be cryptic [hidden] variation within a population so that the physiological induction can be taken over by embryonic inducers." This postulate makes it possible to specify the passage from one level to another and once again recalls the character of transformations specific to each system described by Piaget. In biology the idea has been around for some years that mutations do not always appear ex nihilo but rather that the expression of some mutants is silenced by certain proteins (this phenomenon was illustrated by the discovery of the *Hsp90* protein, a chaperone protein that prevents the expression of some mutations). However, the inhibition of mutant expression can disappear upon environmental change (Rutherfors and Lindquist 1998). This inhibition effectively preserves the mutation by removing it from natural selection in the short term but maintaining it as a tool/resource reserve.

Fourth and finally, "There must be selection for the phenotype." This last point highlights a stumbling block between Piaget's conception of assimilation (in psychology) and Waddington's (in genetics). The latter has nothing to do with Piaget's concept of assimilation. While for both authors assimilation is a process that depends on the "adaptive" nature of the phenotype under consideration, for Waddington this means that the phenotype must be able to be subjected to natural selection, whereas for Piaget adaptation means at most regulation or "re-balancing." Both therefore have a different interpretation of the same idea that *assimilation is an adaptive type of regulatory mechanism*. For Piaget, *regulation* is synonymous with *adaptation* (in the sense of process adaptation), whereas for Waddington it is important to distinguish between the two. Despite this difference, both agree that the mechanism of assimilation involves a significant interaction between the environment and a given organism and that within this interaction it is the organism that is the focus of observation. The disagreement will be over what they consider to be "significant" in the interaction. For Waddington it is the adaptive (selectable) nature of the trait that contributes to the interaction and gives meaning to the relationship between the environment and the organism. For Piaget it is the "interactionist" nature of the relationship between the environment and the organism that is at the origin of adaptation. Whereas the organism and the environment are epistemologically interdependent in the first case, they are ontologically interdependent in the second.

In both cases of adjustment (or regulation), the interaction of the organism with its environment leads to physiological variation. The difficulty, which I mentioned earlier (see chap. 6), is the level at which variation is observed. Can we say that the adjustment visible at the level of the organism is the same as that which occurs at the molecular level? In other words, from the moment one speaks of the interaction of living systems with their environment, does this mean that one has explained the different levels of regulation (which are at the origin of physiological variation)? Is it necessary to identify another criterion?

Let me illustrate this problem: the *Hsp90* protein was identified in a study about canalization processes (see chap. 8, where I return to canalization).

Canalization was first introduced by Waddington (1942) to explain the systematic robustness of developmental trajectories despite the genetic or environmental changes that the organism undergoes during its development (see chap. 3). However, recent experiments have shown that chaperone proteins (such as *Hsp90*) act as "buffering systems" that reduce the sensitivity of the biochemical function of some proteins to minor variations in their structure (variations that may be due to changes in amino acid sequence caused by mutations or changes in configuration due to certain conditions such as temperature elevation). By "buffering" the biochemical activity of these proteins, *Hsp90* appears de facto also to be able to buffer certain developmental processes. At the level of the organism, this protein plays a role of canalization. At the level of the cell, however, its role is no more and no less than that of molecular regulation in the sense of internal control. In other words, at the level of the organism, there seems to be no regulatory mechanism, whereas at the cellular level, there is a molecular regulation that will be at the origin of a physiological variation. In light of this example, it seems that we can speak of a strong pluralism in the explanations of variation, since the same phenomenon can be covered by different explanantia depending on the level of observation. Alternatively, the existence of multiple levels of regulation and plasticity and their mechanisms, and the relative dominance of one over the other, is a subject for future research.

This leads to a first conclusion: the identification of a criterion other than interaction seems necessary to explain the outcome of regulation (physiological variation) at all levels of living organization.

Regulation Driven by a Benchmark

All these considerations oblige me to return to the other meaning of regulation mentioned earlier, which Claude Bernard explained through the notion of "homeostasis" in biology: the specific capacity of living organisms to regulate themselves. In this specific context, the notion of regulation contains, as we have seen, the idea of compensating for the variations of an internal environment, and it integrates what will provide the second characteristic formalized by Canguilhem in his proposed definition of regulation: the reference to the idea of a *criterion* or *benchmark*. Much attention has been paid in biology to the question of functions and, more specifically, to the question of functions that are likely to regulate other functions within organisms. From this point of view, as mentioned above, the reference to regulators in biology is strongly influenced by the use of this term in physics, which allowed the internal elements of a system to be taken into account. Once transferred to biology, the metaphor refers both to regulatory functions within the organism and to regulatory functions within populations. As early as the 1920s, the term *regulation* was used in functional studies. For example, in 1927 George Scarth published an article titled "Stomatal Movement: Its Regulation and Regulatory Role; A Review," in which he refers to the notion of regulation to account for the regular changes that can occur during a particular physiological phenomenon (Scarth 1927). He describes the movement of stomata (a type of plant respiratory organ), their opening and closing according to external conditions. The stomata open when there is an excess of water in the plant to allow water to evaporate; when there is a lack of water, the stomata remain closed to prevent the plant from drying out too quickly (see fig. 6.8). The regulation of stomatal opening is thus dependent on ambient humidity, both for individual plants and, in an evolutionary sense, for many plant populations.

In the context of the analysis of physiological functions, the biologist generally seeks to understand the causes of physiological variation: what makes it possible to observe the transition from one state to another (e.g., what modulates the function of closing or opening stomata in a plant). It also seeks to visualize the possible consequences of the function in relation to the organism as a whole in so far as the function itself can regulate the whole, that is, establish a return to equilibrium (e.g., in the case of stomata, by regulating the plant's transpiration). In functional studies regulation is understood as "the adapted modification of the form or behavior of an organism under changing conditions." Again, the meaning is different from the use of the term in embryology (in the sense of regeneration). The use of the term *regulation* in functional studies tends to bring this concept closer to that of "adaptation," since regulation is understood as "physiological adaptation."

Finally, the term *regulation* is used in ecology in a similar sense to that used in economics: to describe the regulatory component of interspecies relationships. Since the 1960s most discussions in ecology have focused on regulation at the population level. For example, biologists focused on analyzing the role of intraspecific competition in regulating natural populations (Nicholson 1954, 1957). The use of the term was then strongly influenced by the work of Norbert Wiener ([1961] 2019) on cybernetics (systems science[5]) and that of William Ashby (1956) on the theory of control in engineering and biological systems. Both authors view organisms as *regulated systems* that actively maintain or systematically adjust their internal state to counteract the changing conditions to which they are subjected. The form of regulation most discussed in this work is the phenomenon of feedback control, which is found in many homeostatic systems. In these systems regulation is understood as internal and intrinsic to the system under consideration. The proper functioning of the system depends on the activity of the regulatory module, just as any other function of the system is considered necessary for its sustainability. Regulation is thus understood as a structuring and nonaccidental component of the system. The underlying idea is not that of a return to the norm, as was the case with regeneration, but rather the idea of conservation or maintenance of the norm despite disturbances, hence the idea of a benchmark introduced by Canguilhem in his definition.

From an ecological point of view, canalization can also be seen as a form of regulation. But at the same time, regulation no longer necessarily explains physiological variation. On the contrary, it may tend to explain the maintenance of a certain stability. The notion of regulation thus seems to have a certain heuristic flaw, since it does not necessarily explain the physiological variation that can be observed in nature. Regulation seems to be conceived as a structuring component of living systems, but again, if it sometimes explains physiological variation (in the same way that embryonic induction sometimes explains morphological variation), the physiologist cannot use it as an explanans of physiological variation. From a biological point of view, this is its main flaw.

It remains to be understood why biologists often prefer to use the concept of plasticity rather than that of regulation to explain certain physiological variations they observe, particularly depending on whether they are at the cellular or the molecular level.

FROM REGULATION TO PLASTICITY: WHAT ARE THE BENEFITS FOR EXPLAINING VARIATION?

It was not until the early 1960s that questions of regulation began to be investigated at the molecular level. In 1961 the work of François Jacob and Jacques Monod revealed the existence of regulators of gene activity (Jacob and Monod 1961). They analyzed a specific case of regulation at the molecular level: the regulation of lactose in *Escherichia coli* by what they called the lactose operon. This discovery earned them, together with André Lwoff, the Nobel Prize in Physiology or Medicine in 1965. The regulation of the lactose operon genes was the first complex genetic regulatory mechanism to be elucidated, and it is one of the best-known examples of prokaryotic gene regulation. This phenomenon can explain the maintenance of the bacterial growth even in the absence of lactose, the main substrate for enzymatic synthesis, but in the presence of another sugar such as glucose. Depending on the substrate present in the medium, the lactose operon is activated or not (activation in the presence of lactose). In the context of a functional explanation, the regulatory mechanism provided by the lactose operon can be interpreted as a function of the cell that allows it to ensure its growth regardless of the carbon sources available. Regulation at the molecular level (via the lactose operon) allows the proper functioning (and sustainability) of the whole.

The interpretation of this type of regulatory model at the molecular level is slightly different in cases where the regulated genes are not involved in vital functions for the organism under consideration (e.g., cell survival) but in the expression of variable morphological phenotypes depending on environmental factors. There are many examples of this type of regulation (nonessential for survival), both in plants and in certain animal species (shapes and colors of butterflies, colors of grasshoppers, etc.). This type of phenomenon is based on reaction norms that can be represented by curves that relate different phenotypes to variable environments for a given genotype expressing a particular characteristic. This phenomenon, as we have seen, is called phenotypic plasticity. Carl Schlichting and Massimo Pigliucci have shown the correlation between these phenomena of phenotypic plasticity and certain phenomena of gene regulation (Schlichting and Pigliucci 1993). In these cases, it is more difficult to speak of "correct functioning," because how can we know what the benchmark of the regulatory mechanism is and even whether there is one? In other words the criterion of the benchmark (identified by Canguilhem) no longer seems adequate to qualify this type of regulation because it is impossible to say in relation to which internally recognized state that would trigger a reaction this regulation is taking place.

This observation leads to the question of the scale at which regulation is observed. Depending on the scale considered—be it the population, the organism, the cell, or the genome—regulation is not necessarily applied in the same way, and the explanatory role of regulation in these different situations may vary greatly. Will regulation always be able to explain an observed organic physiological variation? Is the analysis of regulation at the molecular level sufficient to explain organic physiological variation that could not be explained at the cellular or organismal level? Unfortunately, it is a far from being that simple.

For example, if we look at an organism as a whole and are interested in its functional regulation, we will compare it with other organisms, whether they belong to the same species or not, and try to show what the normal functioning of that organism is. The appropriate discipline will be physiology. The explanation of this functioning will be based on the analysis of the relations of interaction between the organized parts (which can be seen as the "unstable elements" in Canguilhem's definition of regulation), that is, on a form of internal regulation where any form of *dysregulation* will lead to a pathology or a physiological anomaly. In this type of study, the vocabulary of regulation makes it possible to explain physiological variation by describing a phenomenon of interaction between organized parts, but above all by presupposing the existence of a benchmark, of standard parameters that guarantee the proper functioning of the whole. We can already see that references to the concept of plasticity are rare in this context.

Still at the level of the organism, regulation can also be viewed from a different angle, taking into account the interactions of the organism with its environment. The object of study is no longer simply the organism but the interface between the organism and its environment. This interface, which is dynamic, is the site of regulation. In this interactionist perspective, the elements involved in the interface can be physicochemical data of the environment (temperature, osmotic pressure, pH, etc.) as well as other organisms (plants or animals, of the same or different species) that share the same environment as the organism in question (ecosystems). In this type of regulation, the reference parameters considered (the benchmark against which the regulation is determined) may be any identifiable stable state.

From an evolutionary point of view, therefore, regulation depends on adaptive capacity, which in turn depends on selective pressures. These selective pressures, which can be regarded as invariants, then provide the benchmarks for regulation. This view was further developed by Richard Dawkins with the concept of the "extended phenotype" (Dawkins [1982] 2016), according to which the phenotype is not *limited* to the result of gene expression through biological processes such as protein synthesis or tissue growth but must be *extended* to all resulting manifestations, starting with the behavior of the organism in its environment (e.g., bird's nest, spider silk, etc.) and even beyond its contours. Contrary to the traditional approach, Dawkins's conception implies considering a more "extensive" system of regulation than that usually identified in the description of the organism with its environment. For him, regulation in a strong sense explains the observed physiological variation since we are describing both the site of interaction (the extended phenotype) and the steady states (the constant selection pressures). Again, there is no obligatory reference to the notion of plasticity.

Dawkins's view has been the subject of an alternative hypothesis (though not as successful as his) that is interesting for understanding the potential problems of using the notion of regulation in an interactionist approach to explaining physiological variation. This is the *niche construction* hypothesis, put forward in 2003 by the biologists F. John Odling-Smee, Kevin Laland, and Marcus Feldman in a book of the same name. According to its proponents, by interacting with its environment, an organism can modify the selective pressures acting on it by progressively modifying its environment (e.g., beaver dams, anthills, etc., allowing for a gradual environmental change). The process is thus described as parallel to natural selection. The idea behind this constructionist program is therefore that it is no longer possible to identify an *invariant* in the interaction that the organism maintains with its environment, because, contrary to the traditional evolutionary perspective, selective pressures are no longer considered to be constants or stable pressures. Thus, in this perspective, there is no longer a stable state to characterize the regulatory mechanism. Despite the problems it raises (in particular the ability to build models without using invariants), this approach raises the question of an adequate definition of regulation in a new way. And perhaps this is also the reason why *regulation* appears less than *plasticity* in this context. For this reason, the term *plasticity* (rather than *regulation*) is often used by authors who adhere to the idea of niche construction (e.g., Donohue 2005), in particular to explain the origin of physiological or even ecological variation.

In conclusion, it seems that at the level of the organism, as soon as any reference to a benchmark or a stable state disappears in order to describe a mechanism that would be at the origin of variation, the notion of plasticity seems to be preferred to that of regulation.

At the molecular level, the phenomenon is much more complex. Of course, it has already been observed that a similar situation arises in the case of phenotypic plasticity: it becomes difficult to say whether the benchmark is given by the regulatory mechanism or by the limits of plasticity. However, the commonly used notion of "norm of reaction" seems to reintroduce a benchmark. From this perspective, the generation of physiological variation seems to be explained within the system by what we can continue to call regulation. The niche construction model deprives the biologist of any reference point or stable state and thus prevents him from thinking of the genesis of physiological variation as the result of a regulation. In a way it forces the biologist to refer to a subsidiary concept (the concept of plasticity) to explain this variation without reference points. On the other hand, at the molecular level and with the notion of genetic regulation, the benchmark reappears in the form of the norm of reaction. It is the latter that makes it possible to explain the regulation of the system in the face of environmental perturbations and hence the genesis of variation.

At the molecular level, the situation is somewhat different from that at the organismal level, especially since the term *plasticity* is often used to describe forms of regulation that may generate physiological variation but that can also be seen as being determined by a benchmark (in the case of phenotypic plasticity). There is an additional difficulty at the molecular level:

it is more difficult to distinguish physiological variation from developmental variation—habitual variation from "accidental" variation. It seems that whenever the regulatory mechanism is recurrent (in the sense that its action within a system is repeatedly observed), there is little or no reference to the notion of plasticity (except in the case of phenotypic plasticity). On the other hand, when such a mechanism is "accidental" (in the sense that it occurs only exceptionally), reference to the notion of plasticity is much more common in the writings of biologists. One possible interpretation of this difference in usage is that at the molecular level, the idea of "recurrent" regulation corresponds to the most common use of the term *regulation* (with the different criteria mentioned above), whereas the idea of "accidental" regulation would be quite close to the idea of "regeneration," which was used as a synonym for regulation at the beginning of the nineteenth century.

However, if we compare the process of regeneration to that of the norm of reaction on a graph, both processes can be represented by a bell curve, although the orientation of the curve obtained is different (fig. 7.1). In the case of regeneration (left), the process can be represented by an inverted bell curve because any return to the norm implies a return to "type," and any deviation from the type remains undefined, plastic. In such a view, regeneration appears as the norm, where the ball must end up. In the case of the norm of reaction (right), the process can be represented by a bell curve where plasticity is the norm because it is not easy to "stand up" on the top of the hill. In other words, physiological variation is the norm.

Overall, there is a divergence in the use of the notion of plasticity in relation to regulatory processes depending on the level of the organization under consideration. But within a given level, the notion of regulation per se will be considered more or less appropriate to explain the observed physiological variation. This also explains why the notion of regulation does not necessarily

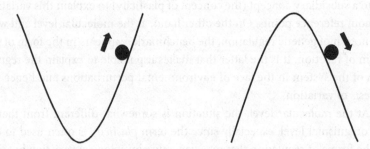

FIGURE 7.1. Two scenarios of the regulation process. (*Left*) Regulation in terms of regeneration; adjustment involves a return to type. (*Right*) Regulation in terms of the norm of reaction, where variation is the norm. Drawing by the author.

allow the biologist to explain physiological variation when she moves to a smaller scale, the cellular or genomic level.

In trying to understand the mechanisms of regulation by plasticity, I have been able to highlight differences in the use of these two notions of induction and regulation. Regulation, like induction, is a mechanism to explain the origin of variation. More specifically, it shows how living systems can be modified as a result of their interaction with the external environment. This modification can be demonstrated once we are able to identify a benchmark, a state that remains stable during the interaction subject to regulation. However, the concept of regulation, even if it allows us to think about the instability inherent in any living system, does not say anything about certain emergent phenomena, unexpected or in permanent instability, that can occur following this type of interaction. Nevertheless, the biologist will at least try to describe these unstable and in many ways elusive phenomena. Sometimes, however, he will have to consider these kinds of phenomena as such, just as he has long been (and continues to be) concerned with describing the extraordinary stability of living things. In this context, the notion of plasticity, which makes it possible to take into account both (a) a norm of reaction at the origin of regular variation and (b) irregular variation as a source of novelty, allows the biologist to take into account mechanisms similar to regulation but outside the benchmarks that delimit the stability criteria assigned to regulation. In this sense, if regulation can be seen as a particular explanation of variation (in the same way as induction was), it seems that the reference to plasticity, understood here as ordered deregulation, allows us to envisage different levels of explanation of variation

General Conclusion on How Plasticity Explains Variation

When Waddington drew the epigenetic landscape, his intention was not simply to describe the process of development but to explain it. What Waddington was trying to show through his landscape, and what many biologists are still trying to understand, are the mechanisms that cause variation during development. Waddington wrote of this depiction: "At the beginning of its journey, development is plastic, and a cell can become many fates. However, as development proceeds, certain decisions cannot be reversed" (epigraph of chap. 6, Waddington ([1957] 2014), 29). This is not a scientific explanation, but rather an intuitive, albeit illuminating, explanation of the developmental process. Since Waddington, many explanations have been proposed to explain the variation between cell types, the combination of which makes up the shape of living things. Thus, embryonic induction explanations describe how cellular tissues differentiate to produce shape variation, and regulatory explanations show how interactions within living systems also cause variation from fixed benchmarks. Both are causal-mechanical explanations. If one of the main aims of biologists is to identify the peculiarities of each of these causal-mechanical explanations of variation, they also try, more or less consciously, to grasp the link that can unite them. Whereas in the past they used metaphors or "powers," biologists now try to identify the different mechanisms responsible for shape change.

However, as biologists have described living phenomena in greater detail, focusing on the molecular level, they have found that understanding phenomena at the cellular or organismal level is not necessarily more precise: it is difficult to simply reduce explanations from one level to another. While some philosophers of biology (such as Dupré 1993) have noted a *disunity* that affects biology as a whole, it seems that explanations of variation are subject to such disunity among biologists. Indeed, in biology, we often find ourselves in situations where the causal-mechanistic mechanisms identified do not allow for a unified explanation of the observed phenomena. For example, a given regulatory mechanism will not help us to understand the presence of polymorphs within a species; an induction phenomenon will not help us to explain the potential of a stem cell to generate different cell types, and so on. It is in these borderline situations, where an explanation by a defined causal-mechanistic mechanism is no longer possible, that biologists usually refer to the concept of plasticity.

Plasticity thus appears as a boundary concept, since it is used when the limits of causal-mechanistic explanations by induction or regulation have been exhausted. It makes it possible, within each boundary, to "make the link" between the different explanations of variation. Does this mean that by invoking the concept of plasticity, the explanation is less scientifically satisfactory? Undoubtedly, if we stick to Hempel's definition. But if we look carefully and empirically at each of the different situations in which biologists invoke the concept of plasticity, we can be more optimistic. For example, phenotypic plasticity describes a phenomenon at the origin of variation that cannot be explained by regulation or induction alone. The same is true of cellular plasticity, which is cannot be better understood by referring to the molecular level.

The very fact that the concept of plasticity can be seen as a boundary concept suggests that the plasticity explanation of variation is constrained to be valid for a boundary region. Thus, the idea of a boundary as applied to the concept of plasticity can be understood in two different ways, depending on how one views the role of plasticity as an explanans of variation. On the one hand, it means that the functional role of the boundary concept is reserved for the region where the biologist has no scientific explanation to offer: she then refers to a concept with a metaphorical connotation to explain the variation that she cannot explain by the usual mechanisms. But this can also mean that there is a boundary region that defies explanation by recurrent mechanisms, although it can be the subject of a proper scientific analysis: the biologist will deliberately refer to the concept of plasticity in order to distinguish this type of explanation from other mechanisms (e.g., with the concepts of developmental plasticity or phenotypic plasticity).

Kitcher, in "Explanatory Unification and the Causal Structure of the World," has suggested that a good explanation is based on a unification of knowledge, and that the more restrictions an argumentative structure imposes on the arguments that instantiate it, the more rigorous it will be regarded (Kitcher 1989). With regard to the first aspect, it seems that the concept of plasticity, as a boundary concept with metaphorical connotations, makes it possible to link the different explanations of variation. With regard to the second aspect, again, the boundary region in which the explanations of variation by plasticity are located leads to the establishment of limiting criteria determined by the shortcomings of the other mechanisms and that therefore make the explanation of variation by plasticity rigorous.

I have also noted that there is pluralism in the strongest sense of the word with respect to explanations of variation: the same phenomenon can be understood by different types of explanation depending on the point of view adopted. In this context, the concept of plasticity, understood in its "active" sense as that which potentially generates a diversity of forms, is a good candidate for integrating explanations of variation when abstracted from a purely evolutionary perspective.

Even if the use of the concept of plasticity is not always perfectly defined, it is very likely that the concept of plasticity constitutes a real "epistemic unit" (Brigandt 2010) when biologists refer to the concept that—sometimes associated with the boundary mechanisms of induction, sometimes with those of regulation—allows them, as it did for Darwin in his time, to better understand and, above all, to progressively provide solid scientific explanations for the morphological and physiological variation observed in the living world.

The concept of (morphogenetic) plasticity allows for different levels of causal generalization (of different scope) depending on what the biologist is trying to explain (e.g., the generation of variation in development). Developmental biologists refer to the concept of morphogenetic plasticity when they wish to specify that an entity (the organism as a whole, a tissue, or even a cell) has the capacity to generate developmental variation. The embryonic

induction process, on the other hand, is a specific level of causal generalization associated with precise criteria. Biologists will always refer to it in
specific situations to indicate that a precise mechanism that meets these induction criteria ([1] interaction between cells with signaling, [2] generation of
cellular variation and [3] existence of a temporal hierarchy in the process) is
the source of developmental variation. The notion of plasticity thus envelops
or encompasses the notion of induction and allows the biologist to account
for a greater number of phenomena (generating developmental variation)
than induction can. From the perspective of a unified understanding of the
causal mechanisms that drive developmental variation, plasticity can be used
to fill in the heuristic gaps of other types of explanations (which apply to a
limited number of mechanisms, such as the induction process). Therefore, if
the notion of plasticity accounts for different levels of causal generalization,
and if embryonic induction is one of these levels, the question is whether we
can identify other levels of causal generalization that explain the generation
of variation in development. In the next section, I address this question by
considering the case of the regulatory process.

Biological Plasticity, an Explanandum
in Evolutionary Process

Darwin did not simply expropriate the phenomena that earlier biologists wished to
account for and supply them with a new explanation; he reformulated the very propositions that required explanation.
ELLIOTT SOBER ([1984] 2014, 135).

In the previous chapters, I have examined the explanans in explanatory
models in biology and in particular where plasticity is used to explain a range
of biological phenomena. In this chapter, I look at other situations where
plasticity appears in biology, where the purpose is precisely to explain what
plasticity is from an evolutionary point of view. I argue that in such situations, plasticity no longer appears as explanans (i.e., what explains) but as
explanandum (i.e., what is explained). The theory of evolution by natural
selection—first formulated by Darwin and later extended by the emergence
of population genetics—has been so successful in all branches of biology
largely because it can explain many characteristics of living things. But what
does it really explain? The diversity of species? Their ability to adapt? How
populations change over time? All these questions remain sensitive and much
debated by specialists. Another question is no exception: does natural selection explain plasticity in the life sciences? The purpose of this chapter is to
consider what the current literature has to say about this and how a fresh look
at the conceptual debate in evolutionary biology through the lens of plasticity
can provide answers to these questions.

I focus first on identifying, in a general way, the type of object that natural
selection seeks to explain and the way in which it does so. I distinguish two
main situations in which biologists tend to think that natural selection explains phenomena of plasticity in living organisms: canalization and adaptation. I examine each process in turn. Starting with canalization—the ability to
develop a constant phenotype independent of environmental and/or genetic
factors—I show how it is often described as the opposite mechanism to phenotypic plasticity and is often explained in terms of natural selection. Then,
I examine the process of adaptation, as this process is sometimes described

in terms of plasticity (as in the phrase "the body adapts plastically to environmental changes"). In this situation, I examine the origins of a possible convergence between adaptation and plasticity. The analysis of the origin of plasticity in evolution in chapter 4 is useful here. It allows me to shed light on the singular status of plasticity, which is certainly explained by natural selection but is not necessarily causally determined by natural selection. I analyze the implications of such a distinction for understanding the challenges posed by plasticity in evolution. Finally, on the basis of these analyses, I conclude on the dual status of the concept of plasticity, sometimes understood as an explanans of variation, sometimes understood as what is explained by natural selection. I show why I think plasticity should be understood as a "boundary concept" in the life sciences and what I think the implications are for biology.

What Does Natural Selection Explain?

In *The Nature of Selection: Evolutionary Theory in Philosophical Focus* ([1984] 2014), Elliott Sober focuses on the explanatory role of natural selection. He shows that what has changed radically since Darwin articulated the theory of natural selection is not so much the fact that a particular new explanation for the evolution of species has been proposed in biology but the fact that a new kind of explanation has become possible thanks to Darwin's reformulation of some biological phenomena. This "new type of explanation" leads biologists not to interpret each observed phenomenon differently but to *rethink and reformulate* the old propositions about the phenomena in a radically new way. For Sober, Darwin is a pioneer precisely because he showed that it was possible to explain the diversity of species without explaining the individual causes of variation. Thus, to explain the distribution of phenotypes of individuals within a population, it is no longer necessary to learn about the singular development of these individuals. To illustrate what is meant by this idea of "type of explanation," Sober refers to Lewontin's distinction between "developmental explanation" and "variational explanation" (Lewontin 1983) and clarifies the distinction by using an analogy. In a third-grade class, all the children can read. We wonder what the explanation is. There are two possible strategies for providing a satisfactory explanation. The first is to mobilize a so-called developmental explanation, which requires looking at each child in the class, one by one, in order to determine her or his personal journey and to identify exactly how she or he managed to learn to read. It is only by looking at all these individual stories together that a proper explanation can be given as to why all the children in that third-grade class can read. The second explanatory strategy proposed by Sober was to mobilize the variational explanation.[1] It does not proceed by aggregating the individual explanations. It explains "all at once" why all the children in a given class can read. How is this possible? Sober's argument is as follows: if, hypothetically, there is a measure that guarantees that no child who is in the second grade and cannot read will be promoted to third grade, then this measure alone will explain what is observed (i.e., that all children in third grade can read). In fact, if (a) we can explain that the passage to the third-grade level is conditioned by a selection made at the end of second grade—a discriminating selection between children who can read and those who cannot yet read—and (b) we can easily explain that children do not lose their acquired reading skills once they reach the third grade,[2] then the explanation that the group of children in the third grade can read is satisfactory. It is not necessary here to analyze the specific learning of each of the third-grade children. However, unlike the developmental explanation, the explanation based on the selection hypothesis—the variational explanation—does not explain why each individual child in the class—Mary, Peter, Paul, or Jane—can read.

The distinction suggested by Sober between a variational and a developmental explanation has been hotly debated since his time. Karen Neander was one of the first to offer a formal—and also one of the most vehement—refutation of Sober's argument in an article titled "Discussion: What Does Natural Selection Explain? Correction to Sober" (Neander 1988). Neander summarizes Sober's argument as follows. While natural selection can explain the frequency of a trait in the population, only development can really explain why an individual has the genotype and phenotype it has. She partially rejects his argument by arguing that while natural selection can explain the frequency of a trait in the population, natural selection can also explain why an individual has a particular trait (contrary to what Sober claims). Her argument is based on the idea that explaining the frequency of a trait in the population is necessary to explain the genotype and phenotype of a trait. Neander believes that the population as a whole plays a much more important role than individual parents in explaining the presence of a trait in the individual. Therefore, according to Neander, it is more relevant to consider the high frequency of traits in the population to which the individual belongs as part of the explanation rather than the singular role of its direct ancestors.

In 1998 the philosopher of biology Denis Walsh defended Sober's position and offered a possible answer to Neander's objection (Walsh 1998). He suggested distinguishing between *traits-type* and *traits-token* (a distinction that Neander does not make clearly and tends to conflate the two), that is, the *trait-type* observed in the population and the *trait-token* observed in an individual (i.e., the occurrence or singular manifestation of a trait in a given individual). According to Walsh,[3] while natural selection explains the presence of traits-type in the population, it does not explain why individual x has developed a particular trait-token.

A few years later, another philosopher of biology, Patrick Forber, also confronted the problem and adopted a more conciliatory position (Forber 2005). He argues that it is possible to escape the theoretical confrontation by distinguishing two different situations that, if confused, are the source of insoluble disagreements. This implies distinguishing two questions: the *role of selection* in explaining the origin and evolution of biological traits, and the *consequences that the previous question may have* in explaining the biological question, Why do individuals have the traits they have?" The first problem, he argues, can be solved by reference to evolutionary theory. The second, however, requires reference to additional extrascientific hypotheses (that often depend on the theoretical positions of the researcher), which I will discuss in more detail below.

Let me return to the first question. What is the role of natural selection in explaining the origin and evolution of biological traits? Forber shows that one of Sober's mistakes is precisely to think that *all* traits are necessarily linked to a single cause. In the case of a trait linked to a single cause (e.g., a single gene or a single allele), selection will not explain *why* an individual has a particular gene or allele. Sober is right to say that selection does not explain the trait, because it is the molecular causes (such as those at the origin of a genetic mutation) that explain the origin of the trait. However, when it comes to explaining traits that result from the action of multiple factors, Forber reminds us that the situation is different. It is not a question of explaining the presence of a particular gene but of explaining that the presence of a combination of factors (genes or alleles combined) affects the expression of the trait. According to Forber, selection becomes essential in such cases because it is through selection, and selection alone, that a particular adaptation (i.e., an appropriate combination of factors for a particular trait) is selected for. Selection therefore explains both the *origin* (genesis) of that particular combination of factors and the *presence* (the fact that it exists and acts) of that particular combination of factors.

Forber concludes, in a spirit of reconciliation, that it is possible to distinguish two points of view that are not entirely incompatible. The first agrees with Sober that explaining changes within a population does not explain the properties of particular individuals (selection does not explain everything); the second recognizes, in part, with Neander that natural selection can also potentially have a creative power in explaining the appearance of particular traits (when these traits are determined by a combination of factors).

If the first question—what role does natural selection play in explaining the origin and evolution of biological traits?—no longer seems problematic and can be resolved in a conciliatory way, what about the second question—why do individuals have the traits they have?— Forber argues that Sober and Neander are choosing different metaphysical and philosophical positions, and for authors who adopt an "explanatory preemption commitment" developmental explanations will always prevail over any variational explanation. Authors who adopt this kind of metaphysical stance (e.g., Sober or Walsh), variational explanations of the origin of a trait will, by definition, never be relevant to explaining why a particular individual has a particular trait. The explanation of origin, from this point of view, must always be reduced to an analysis of ontogeny.[4]

According to Forber, the debate over the primacy of development over selection or selection over development in biological explanation is mainly a reflection of divergent metaphysical positions. On one side are the "developmentalists" (for whom development takes precedence in the explanation), and on the other side are the "selectionists" (for whom natural selection prevails).

What Forber interprets as a "metaphysical assumption" or a "philosophical commitment" on the part of the various authors can also appear, in a less clear-cut way, as a will on the part of some theorists (in this case philosophers, whose approach is undoubtedly very different from that of many biologists) to highlight the *plurality of explanations* for the appearance of traits in living organisms. Forber does not seem to accept pluralism and seems to think that the choice of explanation is always metaphysical.

However, it is reasonable to assume that Sober would not a priori adopt a developmentalist explanation for the origin of a trait but rather would seek to highlight the diversity of possible explanations for the presence of a trait, and this is closer to what biologists do.

Biologists sometimes focus on explaining the presence of the trait and sometimes on explaining the presence of the trait in a population or on explaining the origin of the same trait. These different approaches do not mean that one explanation is better than another but that there may be a variety of explanations for the same general question (e.g., why does a particular individual have a particular trait?).

Beyond this dispute, theorists seem to forget Sober's starting point, which in our case sheds new light on the problem. Let us return to the reasons that led Sober to propose a distinction between developmental and evolutionary explanations. His aim was to explain the upheaval in biology caused by Darwin's theory of natural selection. For Sober, the theory of natural selection not only explained the diversity of species or variation within populations but also showed how this idea of variation introduced a new explanatory principle (Sober [1984] 2014, 135). This part of the problem seems to be neglected in the objections to Sober (this is the case with Neander, but also with Forber).

The debate seems to be based on a common objection: while the developmental explanation can explain why a particular individual possesses a particular trait, this type of explanation can itself be subsumed under a more fundamental and all-encompassing type of explanation, namely the evolutionary(ist) explanation. Put differently and more formally, since natural selection explains development, all explanations that rely on natural selection must also cover (or overlap with) developmental explanations.

To put it more directly, it seems that, on the one hand, most theorists (including Neander and Forber) seem to agree without hesitation with Dobzhansky's aphorism that "nothing in biology makes sense except in the light of evolution" (Dobzhansky [1973] 2013) and seem to consider that any reference to a developmental explanation to explain the appearance of a trait is based on a metaphysical bias; on the other hand, few theorists (including Sober but also Lewontin), while agreeing with the general idea contained in Dobzhansky's formula, will attempt to understand the specificity of the evolutionist explanation and its relation to other types of explanation. But perhaps there is another way.

Dobzhansky's formula, now almost universally accepted,[5] has become almost a dictum: it seems undeniable that the theory of natural selection (Darwin's theory of evolution) has upset the way biologists explain the phenomena of living things and has captured their full attention. But this idea, which is now generally accepted, adds to another of the consequences that immediately followed the introduction of the theory of natural selection into biology: namely, the major upheavals that the Darwinian theory of evolution introduced into the disciplinary structure of biology itself. Far from underestimating the importance of the theory of natural selection for understanding and explaining living things, Lewontin, Sober, and Walsh try to show how the new type of explanation offered by natural selection could be articulated with (and often even contradict) preexisting explanations, which did not disappear but had to be partially reformulated. For example, if we take the question of species diversity in nature, once we take into account the explanatory importance of natural selection for understanding the maintenance of species diversity (the fact that natural selection can both test and predict the maintenance of this diversity), can we say that everything has been explained in this respect? Does natural selection also predict the differences that may exist between the types of shapes observed and the way they are generated within a species? In fact, biologists obviously continue to offer explanations for various phenomena in biology. They all place themselves within the general framework of the theory of evolution, but they do not necessarily explain the phenomenon they are trying to explain on the basis of the evolutionary explanation.

Thus, there will always be several possible levels of explanation for any observed phenomenon. The main difficulty will be in correctly identifying these different levels. Another difficulty will be to understand how the different levels of explanation involved should be articulated or integrated with each other. For Neander there will be no need to integrate the explanations because she believes that the individual development of the trait depends on the evolution of the trait in the population. For Forber the presence of the trait is probably based on a combination of factors that are the result of natural selection. But what would we do if we had to integrate the explanations?

Take Sober's example of schoolchildren learning to read. Sober uses this example to emphasize the difference between an evolutionary explanation and a developmental explanation. He does not mention the fact that what the selection actually explains is mainly "the passage of children who can read in third grade" and not "the fact that they can read." Nevertheless, let me try to show how these two features might be distinguished. It is indeed easy to imagine that selection by reading test is modulated by certain factors. For example, the ability to answer the reading test will not only depend on whether children *can* read or not, but also on their ability to *answer* the test questions thanks to their understanding of the test *modalities*. This second aspect is based on developmental characteristics that differ from one individual to another. Therefore, the *understanding* of the reading test exercise can "modulate" the selection results from what was expected. On the basis of this observation, we can imagine that not all children who can really read will necessarily be promoted to the next grade[6] (e.g., Inhelder and Piaget [1959] 1991). The selection may have retained the children who can read, but it will not necessarily have selected *all* the children who can read at the end of the second grade.

This observation raises the question of whether selection explains not only the transition to the upper grade but also the modulation (the fact that some children are able to answer the test) and, ultimately, the fact that third graders can read. In other words, in the case described above, is a developmental explanation needed to explain that third graders can read, or is a variational explanation sufficient? Sober's example, as transformed here, not only makes it possible to note the existence of different types of explanation but also shows how these different types of explanations can meet. I will return to Sober's example later.

Meanwhile, biologists seem to find themselves in a situation similar to that described in the example of "modulated selection" when they try to understand the role of natural selection in explaining the phenomena of plasticity. Plasticity, considered as the reverse phenomenon of canalization, appears as a process that exerts a certain bias on the trait under selection. Therefore, the question will be whether the analysis of plasticity mechanisms can be useful (even essential) to understand if (and how) the different types of explanations in biology can be complementary and/or integrated. However, by showing that it is possible to identify certain forms of articulation between different types of explanations of plasticity, we can also, in a second step, justify the validity of Sober's argument in favor of a distinction between different types of explanations.

When trying to identify different types of "plastic" mechanisms capable of modulating the expression of a trait subject to natural selection, two examples come to mind: first, the phenomenon of *canalization* (which can be defined, in a first approximation, as the process leading to the conservation of certain developmental mechanisms independent of phenotypic or genotypic expression); and second, the phenomenon of adaptation (which can be defined, at first glance, as the process leading to the modulation of a trait according to environmental perturbations). The question I want to ask is, What is the role of natural selection in explaining these two mechanisms of canalization and adaptation? By examining these two examples, I will then be able to better understand the role of natural selection in explaining plasticity. Another way of framing the problem will be to ask, What kind of explanation would show that natural selection can favor the presence of certain mechanisms in the population rather than particular genes or alleles?

As in the previous chapter, where I was able to highlight the fact that it may be necessary for the biologist to consider in an integrated way different types of explanations to describe variation and that the concept of plasticity allows us to reflect in an integrated way on different types of mechanisms, in this chapter I will try to highlight how the concept of plasticity can allow us to think about the different processes that can be explained in a unified way by natural selection. These processes are all characterized by the fact that, unlike traits that are fixed or stable over time, they modulate the action of natural selection and can only be understood through their relationship with natural selection.

Canalization Explained Through the Lens of Evolution

HOW DOES NATURAL SELECTION EXPLAIN CANALIZATION?

In biology, canalization is primarily seen as a concept of genetics (Scharloo 1991). Waddington first used the term in 1942 (see chap. 3). He distinguished between "genetic canalization" and "environmental canalization." What is commonly called canalization today corresponds to what Waddington called genetic canalization in 1942.[7] It is now generally defined as the "intrinsic robustness" of developmental processes to genetic or environmental perturbations (Wagner, Booth, and Bagheri-Chaichian 1997).

The phenomenon describes the fact that despite variations during the developmental process, the phenotypic outcome generally remains constant. While most definitions of canalization focus on the developmental process, paradoxically canalization seems to have received much more attention in the evolutionary literature than in developmental biology (Scharloo 1991). This disciplinary focus is due in large part to the special attention evolutionists have given to Waddington's experiments with *Drosophila*. In 1942 Waddington investigated the question of "environmentally induced phenotypes." He showed how the apparent Lamarckian processes suggesting the inheritance of acquired traits could be perfectly explained by neo-Darwinian principles. He observed that the wings of some *Drosophila* became deformed after being exposed to intense heat. When he then crossed the deformed insects with each other and exposed their offspring to high heat again for several generations, he found that some insects with deformed wings appeared without even having been subjected to this treatment individually (Waddington 1942). A trait that initially appeared to have been induced by the environment (an acquired trait in the Lamarckian sense) appeared to have been *genetically* transmitted to the offspring. Waddington called this phenomenon genetic assimilation. It is in this very context (that I have already spoken of; see chap. 3 and 7) that he developed the concept of canalization, which he defined as follows: "Developmental reactions 'as they occur in organisms submitted to natural selection' . . . are adjusted so as to bring about one definite end-result regardless of minor variations in conditions during the course of the reaction" (Waddington 1942, 563).

Waddington's interpretation is that for developmental responses to be tuned, *Drosophila* genotypes must necessarily differ in their sensitivity to environmental influences. Waddington believed that *Drosophila* genotypes differed in their "degree of canalization," with some genotypes less easily diverted into aberrant development than others. Waddington's artificial selection in favor of insects with modified wings promotes the accumulation of alleles that ultimately help to canalize development (i.e., direct it in a specific way) toward the aberrant developmental pathway. The more these alleles accumulate, the less direct environmental stimulus is needed to produce the new phenotype: "modified wings."

Ken Bateman in 1959, and then Willem Scharloo in 1991, showed through various experiments (but always on the fruit fly) that genetic assimilation never occurs in inbred populations, that is, in populations where genetic variation is low. This discovery allowed them to confirm Waddington's Darwinian hypothesis on the canalization phenomenon, according to which a normal *Drosophila* population accumulates many small mutations in its genome, mutations that are silenced, but when the environmental conditions (internal or external) change beyond certain limits, these canalization mechanisms stop and allow some of these cryptic mutations to express themselves, providing a collection of new traits (Futuyma 2005). When one of these traits is naturally (or artificially) selected, it is "fixed" in the genome by repeated crosses of individuals that tend to express it.

It is important to remember that Waddington's work on canalization was part of a wider consideration of how the organism develops from the fertilized egg (see Nicoglou 2018). Waddington's interest in development stems more from his interest in recent discoveries in genetics than from any particular interest in the process of ontogeny itself. But he will interpret in an original way the collection of genes carried by the organism and the interactions between these genes (and with the environment) as a "developmental system" at the origin of the formation of the phenotype. Waddington will explain many phenotypic characteristics in terms of the properties of this developmental system rather than the presence of particular alleles. He saw the canalization process as one of the properties that allows the developmental system to produce similar phenotypic results from different initial genetic information. For him, phenotypic uniformity does not prove the underlying genetic uniformity (Rendel 1967; De Visser et al. 2003).

Waddington will justify his theoretical hypotheses with the model of the "epigenetic landscape" (already mentioned in chap. 7). In this model, the "landscape" is a representation of development in the form of a system. The system parameters are determined by genetic loci, and the space is defined as a possible set of phenotypic states. Space, for a given landscape of phenotypic states, is represented as a surface on which each point represents a particular phenotype (see fig. 8.1). Genetic parameters are represented as "pegs" that pull and stretch the surface until they determine its contours. Epistatic relationships[8] between genetic loci are represented by links between threads through which the loci "pull" the surface (see fig. 8.1, bottom). The development of the organism is then represented by the trajectory of a ball on the surface passing through a series of phenotypic states, a ball that prefers to roll in the hollows of the mountainous landscape from conception to death (see fig. 8.1, top, and previous chap.).

Waddington uses this now famous representation of the epigenetic landscape to explain that a change introduced at one genetic locus will affect the

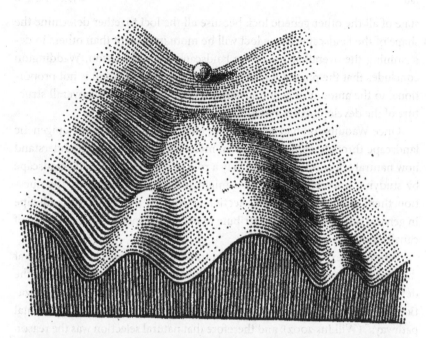

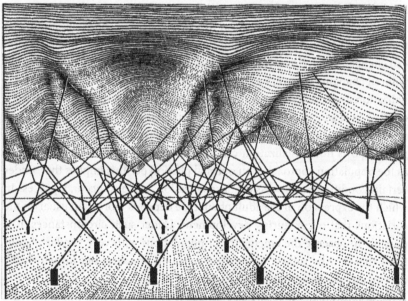

FIGURE 8.1. (*Top*) Waddington Epigenetic Landscape (Waddington 1957, fig. 4, p. 29). (*Bottom*) Foundations of Waddington Epigenetic Landscape (Waddington 1957, fig. 5, p. 36).

state of all the other genetic loci, because all the loci together determine the shape of the landscape. Some loci will be more important than others in determining the overall shape of the landscape. For this reason, Waddington concludes that the phenotypic effect of a genetic modification is not proportional to the amount of genomic variation but depends on the overall structure of the developmental system under consideration.

Once Waddington had succeeded in altering the curves of the epigenetic landscape through *artificial* selection, many biologists sought to understand how natural selection might also play a role in shaping the genetic landscape by studying how *natural* selection might influence canalization. The question, then, was not how natural selection might alter the epigenetic landscape in general but how it might contribute to the appearance or disappearance of canalization phenomena.

Numerous experiments on different traits and organisms have shown that the phenomenon of developmental canalization is a common feature in the development of living organisms. It became clear to biologists that canalization was necessarily the result of natural selection acting on "developmental pathways" (Wilkins 2002), and therefore that natural selection was the reason for the high frequency of canalization phenomena in the living world.

One interesting example is Sean Rice's 1998 article "The Evolution of Canalization and the Breaking of Von Baer's Laws."[9] Rice shows that evolution can change the developmental processes underlying the expression of a trait without changing the average expression of the trait itself. In other words, canalization itself can evolve, since an unchanged trait is ultimately more "buffered" against genetic or developmental variations when it is canalized. From a "weakly canalized" state, the trait can become "highly canalized." Rice's model shows how canalization can change (toward increased canalization) as the population approaches an optimal phenotype. This implies that, under certain conditions, a process of "decanalization" can also occur, a process in which the expression of the adult phenotype depends more than usual on the variations that occur during development. Under such circumstances, it is possible to observe the emergence of populations with very similar phenotypes that may be subject to identical selection pressures but for which the developmental patterns diverge sharply.

This and similar studies (e.g., Flatt 2005) have shown how population genetics can be used not only to model but also to explain the evolution of canalization, that is, to explain the strengthening or weakening of certain developmental processes that can affect phenotypes by acting on gene expression. In this case, if natural selection has many important effects on the evolution of canalization, it remains to be seen whether the identification of these effects is sufficient to say that natural selection "explains" not only the evolution of canalization but canalization itself. In other words, if natural selection is indeed a cause of the evolution of canalization, is it the primary cause of canalization itself?

Despite numerous studies and observations that have taught us a great deal about the evolution of canalization, the question of defining the boundaries of canalization (when it appears or disappears) remains a problem for biologists. There are competing assumptions about the nature of the process, with canalization sometimes seen as causally determined by specific alleles or genetic factors and sometimes as an emergent property of development. These two hypotheses, as you can imagine, do not mix well. Moreover, this problem brings us back to the question of the relationship between canalization and plasticity, since understanding what canalization is from a theoretical point of view often involves understanding what plasticity is, since some biologists see plasticity as the opposite of canalization[10] (in the sense that canalization is the process that leads to the expression of the same phenotype for several different genotypes, whereas plasticity is the process that leads to the expression of different phenotypes from the same genotype).

FROM CANALIZATION TO PLASTICITY

Although the question of the evolution of canalization from a genetic point of view is still very little studied compared to the evolution of genetic mutations (to take just one example), many studies have tried to show that canalization can be selected for and thus explained from an evolutionary point of view like any other trait (e.g., Wagner, Booth, and Bagheri-Chaichian 1997; Eshel and Matessi 1998; Rice 1998). However, biologists are far from unanimous about this possibility. For some of them, canalization should not be understood as a trait but as an emergent property of developmental systems (Siegal and Bergman 2002), the evolution of which would not depend on the direct action of natural selection on the alleles of canalization but rather on its indirect action on functional traits (Wagner and Altenberg 1996). Canalization would, in a sense, be seen as a by-product of selection on traits.

One can see the connection that can be made between this view of canalization and some of the evolutionary views of phenotypic plasticity, which consider plasticity as a by-product of evolution (see chap. 5). However, if genetic canalization is defined as the expression of a single phenotype from multiple possible genotypes, then phenotypic plasticity (defined as the expression of multiple phenotypes from a given genotype) can be considered the environmental counterpart of genetic canalization.

Because of the possible link between these two concepts and the increasing attention given to the issue of phenotypic plasticity since Bradshaw's 1965 article, some of the theoretical issues raised by the introduction of the concept of phenotypic plasticity into the literature have been linked to the concept of canalization.[11] In the case of canalization, natural selection acts not only on the canalized trait but also on canalization as a specific trait. Moreover, canalization itself seems to play a role in selection. Indeed, as a "developmental property" its effects modulate the action of natural selection on a canalized trait relative to another noncanalized trait. The question then arises, How do we know that natural selection provides a similar explanation for the canalized trait and for canalization? It is on this point that biologists disagree when they claim that canalization should not be understood as a trait, but as a developmental property of organisms that would influence their variability, that is, in the final selection itself.

To fully understand the scale of the problem, it is first necessary to specify the meaning of the distinction between what introduces a modulation and the modulated trait. I go back to the analogy suggested by Sober between the selection of the reading test at the beginning of third grade (which explains why all the children who passed in the third grade can read) and natural selection (which explains the presence of a given trait in the population). If we consider that there is a selection bias that is not anticipated by the teachers, such that a relatively large number of pupils who take the test *do not know how to answer the questions asked in the questionnaire given to them* even though they know how to read. Knowing how to "answer the questions" simply means that second graders understand and are able to answer the typical exercise that they are asked to do in order to check their level of learning to read. Their success in the test therefore depends on two things: their ability to read (which does not vary once acquired) and their understanding at the time of the test of the terms or implicit rules of the test (which may vary). Therefore, some pupils may fail even if they can read, but they cannot pass the test if they cannot read. So we have a situation with four different types of people: S1 pupils, who can read and understand the test (these pupils pass the test and move on to the next grade); S2 pupils, who cannot read or understand the test (these pupils fail the test because even if they have understood the rules of the exercise, they do not have the answer to the questions asked because they cannot read); S3 pupils, who can read and do not understand the test (these pupils fail the test and therefore do not progress to a higher grade even if they can read); and finally S4 pupils, who cannot read and do not understand the test (these pupils fail the test and do not progress).

Selection plays its role well, since no child moves to the next grade without being able to read. The variational explanation holds and explains *why* all the children in the next grade can read. However, looking at this particular case, one might ask what the notion of *selection* really explains. Selection, as we have seen, works well for "reading ability" (since only those who can read move on to the next grade). However, in our example, it turned out that this particular trait, "ability to read in the upper grades," did not depend only on the implementation of this reading ability but also on other adjacent elements or factors, such as the children's ability to understand the test well but also the ability of the test to assess only what it is supposed to assess. If selection explains why children can read at the higher grade level, it also explains why these same children passed the test. Selection seems to explain both phenomena in the same way (which explains why a possible selection bias or factors of selection bias may go unnoticed).

Let's summarize. What can we understand from this analogy to natural selection? We have a situation where a "developmental property"—here "the ability to understand a reading test"—is a property that, although not directly visible in the population (as opposed to a phenotype, such as a child being promoted to the next grade if she or he can read), may play an important, even determining, role in the selection process. This is similar to what is called in biology a canalization process, which, as Waddington pointed out, is never clearly apparent in a population. Waddington had to explain the developmental process behind the canalized trait, based in part on a theoretical model (the epigenetic landscape), in order to show the phenomenon of canalization and then to justify that natural selection explained this particular phenomenon.

The school test, like selection in nature, explains how a result x was obtained. But it does not, strictly speaking, explain abilities, let alone how those abilities arise or are modulated. While selection has many *effects* on both objects (canalization and the canalized trait), it explains the developmental trait of canalization only in a second step, after identification. In other words, one must first go through a developmental explanation that highlights the developmental process that *affects* (or *biases*) trait selection. Thus, it is the developmental explanation that helps to explain canalization before the phenomenon is explained by natural selection.

When Waddington demonstrated the phenomenon of canalization, he explained it in terms of a Darwinian process. It was the demonstration of this process that led geneticists to rely on this phenomenon to show how natural selection can explain a developmental trait (such as canalization) as it explains any other trait. But Waddington had also used a developmental model, the epigenetic landscape, as the basis of his demonstration to show the relevance of the Darwinian process. Yet this developmental part of the explanation for canalization is often neglected.

Thus, even if the developmental explanation can a priori be subsumed under the variational explanation, as Neander or Forber argue, it still seems necessary to rely on a developmental explanation to justify the relevance of the variational explanation. Modulation phenomena have generally been thought to occur within what biologists call a developmental pattern, which in Waddington's account is one of the paths that would be taken in the epigenetic landscape.

Earlier I pointed out that phenotypic plasticity can also be considered as part of the type of phenomena that, like canalization, can modulate the effect of selection on a trait. However, biologists do not always think of phenotypic plasticity as a property of a "developmental pattern." When Bradshaw spoke of plasticity, he did so from the perspective of population genetics and never referred to a developmental pattern. This means that, in general, these phenomena that modulate the effect of selection on a trait cannot be considered properties of "developmental patterns." So how do we define them? Canalization and plasticity, unlike phenotypes, are characterized by their lack of stability during evolution. Unlike developmental stages (e.g., gastrula stage), fixed phenotypic traits (number of *Drosophila* wings), alleles, or genes, they are processes or mechanisms whose expression changes rapidly over time. However, these mechanisms are recurrent and as such can be described as a pattern. They may be under selection, but unlike other mechanisms or processes (such as adaptations, discussed in the next section), they are not always directly visible in populations. To identify them, we must first refer to specific theoretical models that allow us to understand them, which will then allow biologists to explain the role that natural selection may ultimately plays in their maintenance or disappearance within populations over time.

Thus, if the explanation of this type of phenomenon (canalization or phenotypic plasticity) is based on natural selection, it seems necessary to go through another type of explanation, a "developmental explanation" (to use Sober's terminology) or a "causal/mechanical" explanation (to use Schaffner's expression; see chap. 6) if one first wants to identify or define (in the sense of determining the contours) the phenomenon to be explained. The evolutionist explanation will only be relevant once one has provided an explanation of the pattern of the process. Once the phenomenon to be explained has been clearly identified, we can say that natural selection explains the process of modulating the effect of natural selection.

Just as it seemed necessary to go through a developmental explanation to really understand what the process is that modulates selection by the reading test and then to say what selection explains—that (1) the third-grade children can read, and that (2) these children "can understand the reading test"— it is also necessary to go through a developmental (or causal-mechanical) model to understand the basis of canalization (or phenotypic plasticity) in order to say what selection explains—in this case, (1) the presence of the canalized trait in the population and (2) the canalization of the trait itself). So, plasticity, as a process that modulates the effect of natural selection on a trait, can be thought of as being explained by natural selection. However, if natural selection can *explain* plasticity, this does not necessarily mean that it *determines* it.

To clarify this question, the next section examines how biologists and evolutionary theorists understand the role of natural selection in canalization. Can a canalized trait be considered an adaptation? I will show that the difficulty that theorists have in agreeing on an answer to this question stems primarily from disagreements about what they mean by *adaptation*. I will now try to clarify their disagreements by showing how part of the confusion is related to the fact that adaptation itself is sometimes understood in terms of *plasticity*.

Adaptation through the Lens of Evolution

THE CANALIZED TRAIT: AN ADAPTATION?

In *The Nature of Selection*, Sober asks whether a canalized phenotype can be considered an adaptation. This question, it should be noted, is different from asking whether canalization or plasticity are themselves adaptations. Building on Sober's analysis, I will now analyze the origins of the confusion in the literature between the terms *adaptation* and *plasticity*.

Sober comments on the position of the biologist George Christopher Williams (1926–2010), who disagreed with Waddington's (1905–1975) answer to this specific question. Waddington believed that genetic assimilation (the result of canalization) could be seen as an adaptive response of organisms in the population to environmental stress. Conversely, Williams believed that the appearance of a canalized trait was always the result of an inability or failure of the organism to respond adequately to the disturbance. What Waddington saw as an adaptive response was, for Williams, "a simple kind of degenerative evolution." His argument was that traits can evolve because they are subject to selection, but this does not mean that these traits are adaptations. The selection regime that Waddington imposed on *Drosophila* resulted in the fixation of a particular trait—the bithorax—but, according to Williams, this trait was not necessarily an adaptation. Ultimately, much of Williams's criticism of Waddington is based on his use of the term *response*. Williams argues with a striking example, that of the Terror during the French Revolution. Williams argues that some Frenchmen would respond to the Terror by conforming to the Jacobin demand, while others would respond by fleeing the country. Finally, some others would respond by losing their heads. This is where the distinction appears. Decapitation is not the result of a response but of a failure to respond "soon enough or in an effective manner." Williams argues that the same phenomenon can be seen in the flies: some responded adequately to the ether, producing a normal phenotype despite the treatment; others showed an inadequate response. The last survived but could only produce a grossly imperfect phenotype (bithorax) (Williams [1966] 2018, 75).

The analogy between the Terror episode in France and Waddington's modified *Drosophila* allows Williams to show that the deformation of the *Drosophila*'s abdomen caused by an environmental perturbation is no more an adaptation than a Frenchman having his head cut off during the Terror. The modified *Drosophila* is therefore, for Williams, not an adaptation, and its modification is therefore neither subject to nor explained by natural selection (I will return to this point later). It should be noted that Williams's analogy between the Terror and Waddington's modified *Drosophila* has some validity: it is a fact that the life functions of *Drosophila*, and hence its survival, are not affected by Waddington's environmental perturbation; the guillotine, on the other hand, is less accommodating.

Before turning to Sober's conclusion on Williams's argument, I would like to dwell for a moment on a point of detail that raises a difficulty in Williams's interpretation of Waddington's work. Waddington saw genetic assimilation as the adaptive response of organisms in a population to environmental stress. He did not see the bithorax (the canalized trait) as the adaptive response to the environmental disturbance but rather as the genetically assimilated trait, which is the trait once the canalized trait is genetically fixed in the population. This is an important distinction. Williams certainly does not overlook this point in his analysis of Waddington's position, but he does not attach much importance to it. For example, in the beginning of his argument (mentioned above), Williams implicitly suggests that for Waddington adaptation was the trait directly caused by an environmental perturbation. He says, "some of the flies . . . showed an *inadequate response* [to ether]" since they were able to survive within the protected culture bottle but could only produce an imperfect phenotype, the bithorax. Certainly the bithorax phenotype described here is not an adaptation, but this is also different from what Waddington says. Only later does Williams give reasons why he thinks that genetic assimilation is no more an adaptation than a trait that has been passively modified by the environment. In other words, by trying very early in his argument to reject an idea that Waddington never really supported—the idea that the modified trait is an adaptation—Williams seems to overlook the complexity of Waddington's experiment and its theoretical context, starting with the epigenetic landscape model. This is particularly damaging because it is precisely this model that establishes the difference between a canalized trait and a genetically assimilated trait (between what constitutes an accident and what Waddington seeks to describe as the norm). For Williams, Waddington's position loses some of its credibility because canalization appears as a kind of demiurge, ordering the phenotype to behave in one way or another according to environmental changes (see Williams [1966] 2018, 80).

Does Sober interpret Williams's critique correctly without losing sight of the complexity of the model proposed by Waddington? As a means of analysis, Sober chooses to compare two types of adaptation: evolutionary adaptation and ontogenetic adaptation. The adaptation that counts for an evolutionary explanation is obviously that which is the result of natural selection—that is, evolutionary adaptation—based on changes in genetic frequencies. But Sober points out that if the lizard's heart rate, which accelerates with increasing temperature, cannot be considered an evolutionary adaptation, then the regular heart rate observed in humans cannot be considered an evolutionary adaptation either! In fact, the specific trait of "regular heartbeat" expressed by Peter cannot be explained by evolution any more than it can explain why Paul, who has moved up in class, can read. Sober's conception of adaptation, with its two types, is ultimately based on the same distinction he makes between types of explanations: the developmental explanation and the variational (or evolutionary) explanation. According to Sober, evolutionary adaptations are necessarily the result of natural selection even if "no organism does anything at all by way of adapting" if it does not respond specifically to environmental cues. The organisms may be totally static, but as long as they live and die differentially, the net result may be an evolutionary adaptation" (Sober [1984] 2014, 205). This allows Sober to conclude that while he agrees with Williams and Lewontin[12] that all adaptation is the result of natural selection, he believes, unlike his two colleagues, that the reverse proposition is also true: "A is an adaptation for task T in population P if and only if A became prevalent in P because there was selection for A, where the selective advantage of A was due to the fact that A helped perform task T" (Sober [1984] 2014, 208). In other words, while Sober believes that all adaptation is the result of natural selection, he also believes that any trait that has been selected for has been selected for a particular function and can therefore also be considered an adaptation.

What does this interpretation of adaptation suggest as an answer to the question of whether the canalized trait in *Drosophila* is an adaptation? Sober seems to agree with Waddington that natural selection is acting on the bithorax trait, and that the bithorax trait is therefore an adaptation. However, Sober also agrees with Williams that, at the beginning of the experiment, the (as yet unassimilated) bithorax phenotypic, as well as the properties of the genome that cause the "bithorax" phenotype to develop, are not adaptations (which, as I pointed out above, Waddington did not claim). Only later, when the bithorax phenotype becomes advantageous in the specific culture flask can it be considered an adaptation. However, as long as we remain at the level of a laboratory population, Sober is reluctant to speak of adaptation, preferring to speak of "accidental benefit." Nevertheless, he believes that the proposition that "a selected trait is an adaptation" remains valid in the case of Waddington's experiment because the adaptation produced does not depend on a given external environment. It is quite conceivable that the adaptation would remain an adaptation in an environment other than the one from which it was originally selected. Finally, Sober points out the difficulties of using the concept of "adaptation" because he thinks that "a trait may be an adaptation at a given time without being adaptive at that time; and it may be adaptive at a time without being an adaptation then." The problem is that to say that a trait is an adaptation is to make a claim about the cause of its presence (Sober [1984] 2014, 211).

For Sober the distinction between "adaptive" and "adaptation" depends on whether one is looking backward (adaptation) or forward (adaptive). Ultimately, the answer to the question, Is the canalized phenotype an adaptation? will vary from case to case. If for Williams the canalized trait should never be considered an adaptation, for Sober it depends: without being an evolutionary adaptation, one can still speak of an ontogenetic adaptation. But if we move away from Waddington's particular experiment and examine cases of canalization in nature, Sober seems to answer the question in the affirmative. Indeed, at the end of the process of genetic assimilation, the canalized trait can emerge as an evolutionary adaptation once it can be (or has been) subjected to natural selection. Why, then, does this question continue to raise so many difficulties?

While there is no doubt that natural selection explains adaptations, it seems that the reciprocal is more difficult to demonstrate because its demonstration depends on each way in which theorists conceive of an adaptation. In my view, most of the difficulties raised by the question, Does selection explain the canalized trait? (or the related question, Is the canalized trait an adaptation?) are based on long-standing, never clearly resolved confusions about the concept of adaptation. Where do these confusions come from? Let us see how, by clarifying the origin of these confusions, we can demonstrate in a different and more indirect way the validity of Sober's position on the status of adaptation—namely, that it is both a *sign* and an *effect* of natural selection—by including an analysis of plasticity.

ADAPTATION AS PLASTICITY:
THE SOURCES OF A CONFUSION

By focusing on the question of the status of the canalized trait, Williams and Sober inadvertently create a confusion that is problematic for the resolution of their issue. Let me explain. I have shown that the process of canalization can generally be related to another phenomenon, that of phenotypic plasticity, and that in both cases one can observe a modulation of the expression of the trait. This idea of "modulation" found in this type of process leads intuitively to associate the "canalized trait" (modulated) with "adaptation," because in both cases there is the idea of change, of modification, but also of plasticity in the passive sense of "malleability" under the effect of an external constraint (whether environmental or genetic).

However, there is a possible confusion between these three phenomena of phenotypic plasticity (insofar as it is the ability of a genotype to express multiple phenotypes as a function of the environment), canalization (insofar as it is the ability of different genotypes to express the same phenotype as a function of the environment), and adaptation (insofar as it characterizes a trait for a given function that is the result of the action of natural selection on that trait). The locus of this confusion, it is important to note, is not in the characterization of the specific processes I have described above. Rather, the confusion lies in the general meaning of each of these notions of plasticity, canalization, and adaptation because of the proximity that exists between them when they are defined in a very general, nontechnical framework. Adaptation, in a general sense, means "to modify one's mode of functioning in order to interact with the new environment." This general and nontechnical definition does not mentioned natural selection. In a very common way, both in biology and in psychology, *adjustment* or *accommodation* are synonyms of *adaptation*, and "adjustment," "flexibility," and "malleability" are synonyms of plasticity. On this basis, in a very general (and nontechnical) way, plasticity can also be defined as the ability of an organism to adapt to its environment.

It should also be remembered that the French neo-Lamarckian transformists of the nineteenth century, when referring to the concept of plasticity, already thought that this concept should be used in evolutionary terms with a meaning similar to that adopted in physiology. In short, they assimilated under the term *plasticity* two synonyms from common language, namely *adaptation* and *accommodation* (see chap. 4, 6, and 7). The neo-Lamarckians sought to confirm the existence of such empirical links between the two processes of accommodation and adaptation by considering that adaptation, whatever it might be, could become evolutionary in the short or long term. And the semantic proximity that exists in common parlance between the terms *plasticity* and *adaptation* is the source of much confusion, for it suggests that there may also be empirical links between the different processes designated by these terms (in particular between phenotypic plasticity and evolutionary adaptation, but also between canalization and evolutionary adaptation).

However, when Williams attempts to show that a canalized trait is not an adaptation, he does so primarily to show that acquired traits play no role in evolution. According to him, if there are adaptations, visible adjustments at the individual level, even acquired traits (what Sober calls ontogenetic adaptations), then these adaptations do not play a role in evolution, whereas other adaptations (what Sober calls evolutionary adaptations) would play an important role. While Williams's view was predictable, it is questionable whether the basis for his justification (and in particular the meaning of the terms involved in the statement "the canalised trait is not an adaptation") is well defined. Indeed, his position is akin to accepting the premises of the neo-Lamarckians in order to reject them. For it is precisely these premises, which consist in equating adaptation with accommodation, that Williams rejects when he attempts to refute Waddington's arguments (which he regards as evidence of acquired characteristics).

Unfortunately, by arguing in this way Williams tends to impoverish Waddington's thinking. For Waddington's position does not overlap with that of the neo-Lamarckians, and in particular Waddington does not seek to show that acquired traits play any role in evolution. Waddington is simply trying to demonstrate how a trait can be modulated by the environment and how that trait can reappear a second time "spontaneously" without being artificially induced. The problem raised by Williams (the interpretation of the place of natural selection in Waddington's experiment) must therefore be considered in the context of the premises that correspond to it, that is, by first distinguishing the different causes and explanations of the phenomena involved in this historical experiment.

I will therefore reformulate the problem by focusing on what happens at the end of Waddington's experiment, when the artificial selection is complete, that is, when the trait in question has already been canalized or even assimilated. In the case of Waddington's *Drosophila*, the canalized trait (bithorax) is causally determined by the modulation process (by canalization). In fact, the *Drosophila* abdomen would not have been modified if a number of factors had not combined to form a system that caused the change in the "trajectory" of the trait. However, at the end of the experiment, the presence of the trait in the population cannot be explained by anything other than selection for the trait. The bithorax trait is observed in the *Drosophila* population, and this observation is explained by the fact that in the culture flask (and perhaps in nature) the insects have been selected for this trait. The canalization process is the *cause* (in the sense that it *determines*) the change in the trait; natural selection is the *explanation* for the presence of the canalized trait in the population. Canalization does not explain the presence of the trait in the population in the same way as natural selection. In other words, natural selection can be presented as a *noncausal explanation* for the presence of the trait in the population, while canalization is seen as the *cause* of the trait. But the phenomenon of modulation of trait expression can, as I have shown, be explained by different theoretical models (the epigenetic landscape in the case of canalization; norms of reaction in the case of phenotypic plasticity). These models are those that, through the association of different factors, make it possible to explain a modulation of the expression of the trait. These combinations of factors, associated in systems, are thus both the *cause* and the *explanation* of the canalization (or plasticity) that can ultimately be modified. Conversely, the modulated trait is always explained in the same way by selection. It is only when we try to explain selective modulation in a specific way that we arrive at a "developmental explanation."

To clarify this argument, it is worth returning to Sober's example, developed above, of pupils who have learned to read: one can imagine that the ability or inability to understand the reading test is explained by a whole range of factors—developmental, psychological, educational, even cultural—and that it is possible (ideally) to draw a diagram describing the combination of these factors. One could also say that this diagram, or the combination of these factors, is the determining *cause* of the ability (or inability) to understand the test. The presence of the trait "can read" in primary school children, which was "canalized" by this ability to understand the test, is still *explained* by the selection (the reading test given). The modulation (here based on the ability to understand the test) occurred in some way *before* the selection.

This analogy allows me to emphasize that a canalized trait (being able to read while understanding the reading test, or the bithorax trait in *Drosophila*) is clearly an adaptation, just as the trait originally described by Sober (being able to read in a third-grade pupil, or an unmodified abdomen in *Drosophila*) could already be considered an adaptation. The canalized trait is an adaptation, but whether it is canalized or not is not what defines it as an adaptation.

This demonstration shows that, contrary to Williams's argument, accepting the proposition that "natural selection explains the canalized trait," one does not necessarily accept the proposition "natural selection is the cause of the canalized trait." This demonstration also shows that Sober's definition that "to say that a trait is an adaptation is to make a claim about the cause of its presence" (Sober [1984] 2014, 208) is not *sufficient* to characterize an adaptation. Indeed, when Sober writes "to make a claim," he does not necessarily mean that one can (factually) identify that cause but only that one assumes that it exists. Rather, it is thought that a more restrictive definition of adaptation should be adopted, one that is reserved for evolutionary adaptation and that would stipulate that *to say that a trait is an adaptation is to mobilize selection to explain its presence in the population.*

On the other hand, I agree with Sober that, following Waddington's experiment, one can speak of adaptation in terms of the bithorax phenotype. For the important thing is not whether natural selection is at work or not but how we define selection. For Waddington, selection is artificial, but it is a particular mode of selection, and in terms of that particular mode of selection, the offspring of those individuals may or may not "survive" to the next generation. Rather than letting nature "choose" which individuals survive and which die, Waddington does it. The selected trait, the bithorax phenotype, is therefore an adaptation in terms of the way in which it was selected. But an adaptation is only said to be *evolutionary* if *natural* selection is at work (although the boundary between natural and artificial selection is often difficult to define). In this sense, any adaptation must be based on a stable trait, that is, one that can be identified with a certain frequency within a population.

Returning to my original question, is canalization (or plasticity) an adaptation? Given the definition of adaptation that I have just proposed, I can now answer this question in the affirmative. But I have also shown that the term *plasticity*, in the sense most commonly used by biologists, can be compared to that of *adaptation*, and that plasticity does not necessarily correspond to a process of the same kind as canalization. Numerous examples in the contemporary literature illustrate the difficulty many biologists have in making a clear distinction between adaptation and plasticity. As early as 1995, Gotthard and Nylin noted that it was high time to clarify the terminology in order to bring order to the studies of phenotypic plasticity: "Considering the large number of published studies and reviews of (presumably adaptive) phenotypic plasticity, remarkably little attention has been given to the question of what is adaptive plasticity" (Gotthard and Nylin 1995, 4), and "Confused terminology in this field of research may be one reason why the focus of many studies is not as clear as it could be" (1). The confusion highlighted by both authors is that adaptive phenotypic plasticity is sometimes seen as an evolutionary adaptation (like any other trait that has been selected by evolution) and sometimes as a simple benefit (an ontogenetic or physiological adjustment). Let me clarify this.

First, as Gotthard and Nylin point out, most biologists and theorists agree that a trait is adaptive (in the evolutionary sense) for a given function if it can be shown to be the likely result of natural selection acting on that function. On the other hand, Gotthard and Nylin point out that in many studies, a trait is considered "possibly adaptive" (i.e., a possible evolutionary adaptation) simply because it is beneficial. In other studies, being "adaptive" for a particular function simply means that the trait is beneficial for that function as opposed to a nonadaptive or even maladaptive (nonbeneficial) trait. In this context, therefore, a beneficial trait will always be considered "adaptive," and in some cases it may also be considered an "adaptation" (i.e., evolutionary). It is therefore not always clear in which sense *adaptive* is used. To clarify this idea of "adaptive" without an evolutionary meaning, some authors have used another notion—that of inherency or generative entrenchment (Wimsatt 2014)—which implies that some functions expressed by organisms (based on studies of the development of unicellular and multicellular organisms; see Newman 2023) are inherent to the cells from which they developed rather than selected effects. The underlying idea is that the organism cannot develop without these features (examples are gastrulation or neurulation in vertebrate development).

Gotthard and Nylin, on their side, propose to distinguish the use of the adjective *adaptive*, which should be understood as "beneficial for a given function," from the use of the noun *adaptation*, which should be understood as "apparently designed for a function and whose origin is linked to selection." The difference with Sober is that whereas the latter gives equal weight to both terms, seeing them as two sides of the same coin, Gotthard and Nylin seem to give more weight to the term *adaptation* (which becomes a technical term) than to the adjective *adaptive*, which takes a generic turn.

It seems that the confusion in many of the studies mentioned by Gotthard and Nylin has less to do with the meaning of the word *adaptive* than with the meaning of the word *plastic*. Let me illustrate this with an example from a study that builds on Gotthard and Nylin's article and cites it several times (which shows that the process of terminological clarification is far from over, even though studies on adaptive plasticity have multiplied in recent years). The paper is titled "The Evolutionary Ecology of Individual Phenotypic Plasticity in Wild Populations" and was published in 2007 in the prestigious *Journal of Evolutionary Biology*. To briefly summarize the stated aim of the three authors—Nussey, Wilson, and Brommer—is to report on a widespread natural phenomenon, namely, "the ability of individual organisms to alter morphological and life-history traits in response to conditions they experience" (Nussey, Wilson, and Brommer 2007, 831). They call this "phenotypic plasticity." It is clear from the introduction to their article that the authors do not see phenotypic plasticity a priori as an evolutionary adaptation but rather as an individual benefit: "Plasticity allows *an* organism to 'fit' its phenotype to the changeable environment" (831, my emphasis). This phenomenon they mention, which can be considered, for clarity, as a case of individual phenotypic plasticity, is, according to the authors, "fundamental to any population's ability to deal with short-term environmental change." Noting that biology still lacks knowledge about the impact of natural selection but also about the impact of many other causes (such as ecological causes, to take one example) on "variation in life history plasticity," the authors propose an analytical framework based on both the concept of norm of reaction and statistical random regression models to assess "between-individual variation in life history plasticity."

This framework underlies, through its effects, the "population's *response* to the environment" (both phenotypic and genotypic). Thus, the authors consider how natural selection and ecological constraints can modify the population's response to the environment through their effects at the individual level. This study, they say, opens up a new field of research on what they call individual plasticity.

The main problem with this article is the circularity of the argument, especially when plasticity is described as a cause in one place (generalist or intuitive definition of the term) and when it is described as an effect in another (technical definition in the sense of "phenotypic plasticity"). Moreover, it is often difficult to identify precisely what it is a cause or effect of, since plasticity appears to be both a cause and an effect of variation at the genetic, individual, and population levels. Indeed, by the end of the article, it is impossible to be clear whether the authors consider phenotypic plasticity to be an adaptation as defined by Gotthard and Nylin. They seem to argue that it is not an adaptation, since throughout most of the article the term *plasticity* can be replaced by *adjustment* without any change in meaning (Nussey, Wilson, and Brommer 2007, 831).

Consequently, plasticity is sometimes understood as a simple "adjustment" (beneficial or not), sometimes as a specific property or process operating at the level of individuals (it could therefore be understood as an adaptation in this sense). Moreover, the idea that the evolution of plasticity might eventually be predictable is very ambiguous because it suggests that plasticity is indeed an evolutionary adaptation; we could eventually see the effects that natural selection would have on it. This conceptual ambiguity, which is the source of much confusion, reflects the persistence of a recurrent problem in the use of the term *plasticity*: as with the term *adaptation*, it is sometimes difficult to know during the authors' argumentation whether they are sticking to the technical definition initially given or to a more general one. Clarifying whether these phenomena are understood as cause or effect would provide some precision. The difficulty, however, is that plasticity (like adaptation) can be understood as a cause or an effect depending on the scale at which the phenomenon is observed and the context in which one wishes to explain it. This is a subtlety of the concepts of plasticity and adaptation that the biologist should not overlook.

In the previous chapter, I examined the situation where plasticity is understood as an explanans of variation (without much consideration of whether this type of explanation has an evolutionary framework or not). In this chapter, I have tried to understand what explains the phenomenon of plasticity in an evolutionary context. In such a context, it seems necessary to break down the different parts, the different "fragments" (to use an expression of Schaffner's) of the explanation of plasticity; it becomes more than necessary to specify the moments in the explanation when plasticity is to be understood as *a cause of variation* and the moments when it is to be understood as an *effect of evolution*.

This perspective makes it clear that the causal links between the different phenomena described are not always obvious. Common intuition leads us to think that what causes evolutionary adaptation is the same as what causes canalization (or plasticity). This is not the case. Evolutionary adaptation is the result (effect) of natural selection, whereas plasticity is the cause of a modulation of a particular developmental system. But in some cases, (physiological) adaptation can also be understood as the cause of a physiological change following an environmental perturbation, just as phenotypic plasticity can be seen as an effect of natural selection when observed at the population level. Thinking in a unified way about different theoretical contexts risks confusing these different levels of explanation, which are relative to the particular contexts to which they apply. It is therefore not possible to think of an explanation that was valid in one context as being valid in another context without taking into account the change in context. The risk would be, for example, to find studies in contemporary biology that analyze plasticity in an evolutionary context but refer to definitions of the concept similar to those used by embryologists in the nineteenth century.

There is also another major difficulty: there may be an empirical dissociation between explanations and causation.[13] Natural selection may explain the canalized trait, but a developmental explanation is needed to identify the determining cause of canalization. What does the dissociation between causes and explanations tell us? In the particular case of plasticity, just because we know the cause of the plasticity process does not mean that we know what explains plasticity in the life sciences.

Sober had emphasized the importance of reinterpreting certain questions in light of the new theoretical framework provided by the theory of natural selection. Previously, it was not possible to distinguish between explanations and causes of phenomena. The theory of evolution by natural selection provides a new explanatory framework in which it is now possible to reinterpret old questions by multiplying explanations, so to speak.

In this context, the concept of plasticity makes it possible to highlight a hierarchy between the different types of explanation. The shape of this hierarchy would depend on the scale at which living beings are observed. For a given scale of observation, some explanations would be more important than others. Moreover, as I have shown, just because a biological explanation is valid at one scale of organization of living beings, it does not follow that it is valid at any other scale. Therefore, when relying on analyses of plasticity at different scales and from different perspectives, it is important to note that different levels of explanation must be distinguished and that a reductionist approach to plasticity is not possible.

Conclusion: How Does Natural Selection Explain Plasticity

Plasticity explains variation or is explained by selection. Does this mean that plasticity is a boundary concept? At the end of chapter 7, we showed that plasticity can be understood as a boundary concept for two methodological reasons. First, the use of the concept is reserved for the area of the field where the biologist no longer has a scientific explanation to propose. He then refers to a concept with metaphorical connotations to explain the variation that he cannot explain by means of common mechanisms. Second, the use of the concept allows the biologist to distinguish a specific and new way of explaining variation different from induction and regulation.

At the end of this chapter, I have shown that plasticity can again be considered a boundary concept for two epistemological reasons. First, the use of this concept serves to emphasize the need to draw the boundaries of the natural selection explanation. Second, the use of this concept serves to emphasize the need to distinguish between explanations.

First, the study of the canalization process and its links with the process of phenotypic plasticity has allowed me to show that the explanation of the phenomenon can be divided into two parts: on the one hand, what causally determines the phenomenon and relies mainly on developmental explanations (or causal-mechanical explanations if we consider that phenotypic plasticity is not a developmental property), and on the other hand, what explains the existence of the process as a so-called pattern, which, like any adaptation, depends on natural selection. Second, the proximity that persists in the literature between different general definitions of adaptation and plasticity sometimes leads biologists to confuse explanations of the underlying processes with each other. However, by identifying the nature of this proximity, in particular between the notions of adaptation and plasticity, and the confusion it causes, I have shown how this confusion can be avoided by distinguishing processes according to the nature of the explanations associated with them. In a sense, the analysis of the concept of plasticity allows us to think about the question of evolutionary explanation in a broader way. For example, such an analysis allows us to abandon the idea of blurring the boundaries between the different types of explanation by considering that the explanation by natural selection would encompass all other explanations. The relevance of the explanation by natural selection sometimes depends on the evidence for other types of explanation, such as developmental explanations for understanding certain processes (starting with phenotypic plasticity and canalization). Finally, the analysis of the concept of plasticity also forces us to recognize that some explanations that predate the introduction of natural selection now need to be reformulated in an evolutionary context.

The aim of this chapter has been to understand how the introduction of the theory of natural selection has challenged biologists' understanding of plasticity. It seems that one of the conclusions that can be drawn is that while the term was already problematic for biologists trying to describe a phenomenon without some of the old metaphorical and philosophical connotations associated with the term, it became even more problematic with the introduction of the theory of natural selection. Not only does the term retain a strong metaphorical connotation, as it continues to be used in a generalist sense, but its use also tends to become more specific across disciplines, resulting in a proliferation of definitions. Finally, with the introduction of natural selection, the numerous definitions of the term have been joined by various explanations of the phenomenon.

Paradoxically, this proliferation of definitions and explanations of plasticity has not led biologists to stop using the term or to analyze plasticity phenomena in living organisms. Often aware of the plurality of explanations and definitions, they sometimes consider the concept as an "epistemic unit" that allows for the integration of explanations, which I analyze in the next chapter.

9

Plasticity in the Evolutionary Developmental Synthesis
Toward an Integration of Explanations?

[Haeckel] declared that . . . : "Development is now the magic word by means of which
we shall solve the riddles by which we are surrounded, or at least move along the road
towards their solution."

EMANUEL RÁDL (1930)

From the end of the 1970s, a new theoretical turn took place. Some philosophers and biologists (e.g., Gould 1977; Hamburger [1980] 1998; Coleman 1980; Lauder 1982; Wallace 1986) began to criticize the genocentric view of the modern synthesis and brought the question of ontogeny back to the fore, arguing that the analysis of this process could help to explain the variability of interaction. In 1986 Bruce Wallace asked the central question that animated all these theorists: "Can embryologists contribute to an understanding of evolutionary mechanisms?" In the article that addressed this question, Wallace answered in the negative by demonstrating that embryology was not necessary for evolutionary explanations: "Problems concerned with the orderly development of the individual are unrelated to those of the evolution of organisms through time" (Wallace 1986). He highlighted two different views of evolution that he felt should be distinguished. The first view focuses on organisms and sees evolution through the vast set of morphologies that can be found in nature. The second focuses on genetic programs and sees evolution in terms of changes in genetic frequencies (see chap. 8). The two visions seem incommensurable to Wallace, who illustrates this with an analogy: the famous example, familiar to Gestalt psychologists, of an optical illusion in which you sometimes see a beautiful young girl and sometimes an old woman but never both at the same time. For Wallace, the same is true of evolution: it is not possible for the biologist to understand evolution from both a developmental and a genetic point of view. It is therefore not possible to adopt a view conception that combines these two visions.

After the initial success of the modern synthesis in the 1950s, the study of development became less central than it had been in the past. It was not until the 1980s that some theorists and biologists—such as Stephen Jay Gould,

Viktor Hamburger, William Coleman, George Lauder, or Bruce Wallace—turned their attention back to the question of development in evolutionary studies. In chapter 4 I referred to an article by Bruce Wallace in which he wondered whether embryologists could contribute to the understanding of the mechanisms of evolution. He concluded that the explanations provided by the modern synthesis (and more specifically by population genetics) and those of developmental biology were incompatible. Indeed, I have argued in this chapter that in some respects they could be considered incommensurable for Wallace, since population genetics focuses on populations and considers changes in allele frequencies as the locus of evolutionary change, whereas developmental biology focuses on organisms and considers the diversity of morphologies present in nature as the locus of evolutionary change. The role of development in the evolutionary synthesis thus seemed sealed: development could at best explain the diversity of forms in nature. However, despite Wallace's assertions of incompatibility (also known as Wallace's challenge), the study of evolutionary development (whose most famous studies are in the field known as evo-devo[1]) has since emerged and grown dramatically. Does this mean that the incommensurability described earlier has resolved itself? Or does it mean that there is no integration of development into evolutionary studies as such (as some seem to claim)? Or does it mean that there would be no such incommensurability between developmental and evolutionary explanations? There may be no single, unified answer to this question, just as there is no single view in the new field of evo-devo.

One of the reasons for this is simple: although an increasing number of biologists today consider their research to be evo-devo, there are considerable differences in what they think they can ultimately explain within this same disciplinary framework. Some try to explain biological questions based primarily on developmental studies (e.g., the origin of morphological diversity), others try to understand evolutionary phenomena that remain unexplained from a developmental perspective (e.g., the origin of evolutionary novelty), and still others try to propose new models and solutions to old problems that articulate both evolutionary and developmental studies (e.g., modularity[2]).

In such a vibrant context (which, incidentally, makes the present topic even more challenging to explore), the concept of plasticity has gradually emerged as a concept of crucial importance. This is evidenced by the appearance of the entry "phenotypic plasticity" in the book *Keywords and Concepts in Evolutionary Developmental Biology*, in which the editors emphasized that the concept is relevant to both evolutionary and developmental biologists (Hall and Olson 2006).

In the previous two chapters, I showed that "plasticity" in biology can be understood not only as that which explains change—implying that it is the cause/source of the diversity of morphologies among organisms—but also as that which is explained by evolution—implying that phenotypic plasticity, for example, could be selected for like any other biological trait. As this distinction shows, the concept of plasticity can be associated with different types of explanation. It is therefore easy to see why it has quickly become a key concept in the emerging field of evo-devo, which explicitly aims to bring together explanations from both evolutionary and developmental biology. However, as I have shown, biologists do not always clearly define the term *plasticity* when they refer to it. The question, then, is to understand how theorists of a synthetic view of evolution, which tends to integrate development, use the concept of plasticity in their work, and what variations there are in their use. While such a view may appear biased or partial in relation to the whole of evo-devo conceptions, it assumes that "integration" is not always thought of in evo-devo biology, whereas it is claimed by some theorists of the evolutionary developmental synthesis (EDS).[3]

In the first section, I examine the foundations of the different trends within EDS. The second section focuses on one particular theoretical trend within EDS: developmental systems theory (DST), which is often presented as a "radical" conceptual framework that advocates a holistic approach, whereas most approaches in the field have tended to make analytical distinctions. In particular, I examine how DST conceptualized the integration of biological explanations with an emphasis on plasticity. The third section is devoted to an analysis of what I have termed the *theory of developmental plasticity* (TDP), based on the foundational work of the biologist Mary Jane West-Eberhard in her 2003 book *Developmental Plasticity and Evolution*. West-Eberhard argued that the notion of plasticity, which refers to the biological phenomenon of variation associated with the ability of some genotypes to express multiple phenotypes in different environments (what Anthony Bradshaw famously called phenotypic plasticity), should also be seen as a central biological phenomenon. This perspective allowed her to integrate developmental and evolutionary explanations. Finally, in section four, I assess how sixteen EDS theorists in 2009 proposed a new conceptual framework—in which plasticity appeared as a key concept and a key phenomenon—with the aim of establishing an "extended synthesis of evolution" (Pigliucci and Müller 2010).

The Foundations of the Different EDS Programs

There are various "trends" or "programs" of research in EDS. Take, for example, the challenge posed by Wallace in 1986: development cannot satisfactorily shed light on dark areas of the modern evolutionary synthesis because developmental and evolutionary explanations are incommensurable. It is precisely this point that puts two different types of EDS approaches into perspective.

Some biologists, for example, would argue that their work is in no way compatible with work in population genetics. They focus mainly on questions of *how* to understand evolution through a developmental explanation of morphological diversity (e.g., Gilbert, Newman, Müller). They see their own work in a very different light from those who, for example, establish purely evolutionary explanations based essentially on population genetics. Conversely, some EDS biologists believe that Wallace's challenge can be overcome. This is the case of Brian Keith Hall, who, in an article titled "Unlocking the Black Box between Genotype and Phenotype," showed that both disciplines—developmental biology and evolutionary biology—are based on identical foundations and therefore "neither developmental nor evolutionary change [could] be explained by genes alone" (Hall 2003, 220). Hall sought to identify a common basis for these two types of change not just by looking at genes or phenotypes alone but by looking at what lies between the two, namely, the developmental process. He criticized the modern synthesis for neglecting development because the framework treated it as a "black box." He rejected the argument of incommensurability between developmental and evolutionary explanations by suggesting that "opening the black box of development" and understanding how it works would allow evolution to be understood differently. In his opinion, we should stop looking only at what goes on outside the box (i.e., incoming signals—genes—or outputs—phenotypes) and start looking inside the box (i.e., the mechanisms of developmental processes).

The opposition between Wallace and Hall highlights two different research programs among EDS biologists, which in turn depends on how they see the relationship between developmental and evolutionary explanations. This means that some biologists may think, like Wallace, that it is not possible to unite developmental and evolutionary explanations but that they are still doing EDS research when, for example, they offer evolutionary explanations to explain developmental processes. Others, like Hall, may believe that evolutionary explanations can be reinterpreted in the light of some developmental observations. These two theoretical orientations help to characterize a first difference within the field of evo-devo, which is based on an epistemological opposition that allows for *two different ways of understanding* biological phenomena within the field: on the one hand, "interdisciplinarians," like Wallace, think that development can help explain evolutionary phenomena, and on the other hand, "synthesizers," like Hall, think that the Modern Synthesis is incomplete until it takes into account the role of development in it.

Differences between research programs in EDS are not limited to differences in biologists' epistemological assumptions. In fact, few biologists are aware of the extent to which they integrate developmental and evolutionary explanations in their approaches. For the most part, evo-devo biologists tend to be in a grey area between the two extremes, or at least they do not state their position clearly. Even for Hall, his main interest is not in overturning the modern synthesis but in understanding how developmental mechanisms are mobilized to understand the diversity of morphologies (or phenotypes).

Differences in the ultimate purpose of research highlight a source of variation in research programs in evo-devo. This has to do with the methodological orientation of the biologists' works: the difference in their purpose will lead them to prefer one method over another to choose the kind of data they collect. In this perspective, Gerd Müller and Stuart Newman brought to light three different programs (with different methodological orientations) within the field of evo-devo: "the morphology and systematic program" (which, in their view, seeks to analyze the mechanisms that generate morphological novelty), "the gene regulation program" (which seeks to analyze the mechanisms of innovation at the genetic level), and "the epigenetic program" (which seeks to analyze the developmental process itself and its mechanisms)—both Müller and Newman's works fall within this last program) (Müller and Newman 2005). For example, the biologist Sean Carroll, whose work belongs to the second research program—the gene regulation program—suggests that same genes, or what he calls the developmental genetic tool kit, can be reused many times and in different ways throughout the developmental process; according to him, morphological diversity can be fully explained by the combination and/or differential regulation of the developmental genetic tool kit alone (Carroll, Grenier, and Weatherbee [2001] 2013; Carroll 2006). This gene regulation program elaborated the famous theoretical model of the homeobox genes. On the other hand, Newman and Müller have developed the epigenetic program, in which they refer to a different methodology to account for morphological diversity. For example, they take into account the physical or generic properties of cells or tissues of a developing organism to understand how variation (and therefore diversity) is generated during developmental processes. Two different methodologies can therefore be at the origin of two different evo-devo research programs.

Finally, the different research programs and studies in evo-devo may rely on specific associations between different types of epistemological approaches (i.e., disciplinary interaction or synthesis between development and evolution) and different methodological options (i.e., the type of data they choose to collect). For example, among evo-devo biologists who believe that developmental explanations can have the same foundations as evolutionary explanations, some consider genetic data to explain morphological diversity, while others analyze ecological distributions of populations to explain such a morphological diversity. In other words, some of these biologists may think that developmental and evolutionary explanations can be unified by referring only to causal-mechanistic explanations, while others may think that causal-mechanistic explanations need to be combined with statistical explanations (which are relevant when the approach is based on an ecological distribution of morphologies). Several combinations are possible, depending on the theoretical and methodological orientations of individual research programs. This adds to the diversity that already exists within the evo-devo field among biologists.

Moreover, trends in EDS have sometimes been driven by factors that are even more difficult to identify and whose identification requires an examination of some of the historical-theoretical aspects of the field (i.e., even before the influential discussions in the mid-1980s that triggered the most recent wave of interest in evo-devo). For example, when Stephen Jay Gould published his book *Ontogeny and Phylogeny* in 1977 (now considered a key text in the history of evo-devo, see Amundson 2005; Laubichler and Maienschein 2009), he did so initially as a reaction against the nineteenth-century propagandist Ernst Haeckel to criticize the descriptive methods Haeckel employed and to reject the controversial speculations he used to support his argument that "ontogeny recapitulates phylogeny." In reacting so strongly a century later, he was probably trying to discuss his contemporaries' assumptions about the role of development in evolutionary theory, which were based on Haeckel's views. However, too many evo-devo biologists remember only the "return of development" to the study of evolution, forgetting that the original reason for the emergence of the first modern EDS program was precisely the critique of Haeckel's speculations. The book was primarily a way for Gould (and others, such as Hamburger, who were developing similar views at the same time) to highlight the limits of the explanatory power of the developmental theory of evolution as Haeckel had formulated it (Laubichler and Maienschein 2009). Today, when commentators revisit this history, they tend to focus on the response of developmental biologists in the early 1980s to what they described as "the limits of the explanatory power of the adaptationist program,"[4] which was one of the pillars of the modern synthesis. Ultimately, by overlooking the main reason for the birth of the modern evo-devo (i.e., Gould's rejection of Haeckel's recapitulation theory), many commentators miss both what was behind Haeckel's success (before he was rejected) and the reason why Gould wrote *Ontogeny and Phylogeny* in the first place. Gould needed to show that it was possible to look beyond Haeckel's view and thus put an end to Haeckel's view in the first place. Indeed, Haeckel appeared to be one of the first naturalists to propose (avant la lettre) a kind of EDS approach, combining questions of both evolutionary biology (phylogeny) and embryology (ontogeny) (Laubichler and Maienschein 2009). Although Haeckel's approach was far removed from modern notions of evo-devo, Haeckel already gave embryology a central place in his own interpretation of phylogeny. The biologist and theorist Emanuel Rádl, in his 1930 book *The History of Biological Theories*, emphasized this when he quoted Haeckel as saying, "Development is now the magic word by means of which we shall solve the riddles by which we are surrounded, or at least move along the road towards their solution." (Haeckel quoted in Rádl 1930, 126–27). The question remains, Are we not still following the same path that Haeckel opened up more than a century ago?

More than twenty years after Rádl's book, in 1955, Jane Oppenheimer (1911–1996)—probably one of the first historians of the field and a critic of Haeckel long before Gould—was astonished that the embryologists of Haeckel's time had accepted his claims and theories even though he had never been a practicing embryologist (Oppenheimer 1955a, 16). Oppenheimer's observation shows that biology is in fact never free from a number of a priori considerations. And in a sense, biology today is no more free from this risk than it was in Haeckel's time.

And indeed, evo-devo biologists or EDS theorists may forget that Gould was not so much trying to refute Haeckel's position but Haeckel's a priori idea that any evolutionary explanation should be based on an understanding of development. The point here is not to undermine the idea that development plays a crucial role in understanding certain evolutionary phenomena. Rather, it is simply a warning against confusing *a methodological approach* with *an underlying metaphysical assumption*, since such assumptions can substitute for demonstration or preemptively reject the method employed. Indeed, the danger would be to assume that development is, by definition, more suitable for evolutionary explanations than any other kind of explanation. Under these conditions, the idea that development is more likely to lead to a better understanding of biological phenomena is as flawed as Haeckel's theories, because it is based on an unjustified assumption. A different approach would be to consider that development can be a source of explanation among others. In any case, I think I have sufficiently shown here that some of the differences in EDS programs may also be based on these differences in metaphysical assumptions, according to the author's ideas.

The assumption that developmental explanations are superior to all other types of explanation is not shared by the majority of evo-devo biologists. However, the identification of this type of assumption serves to emphasize that there can be enormous differences within the theoretical programs of EDS (Müller and Newman 2005, 497).

To take an overarching view, it is possible to describe EDS as a *patchwork* of programs, reflecting methodological differences between authors, that in turn reflects the different disciplinary influences of the biologists involved. For this reason, Laubichler and Maienschein's 2009 definition of evo-devo seems most eloquent in capturing the unity of the field, which persists despite significant differences in EDS trends: "For clarity, I am defining 'evo-devo' rather broadly, as a synthetic paradigm or research program in which developmental biology is combined in some ways with both genetics and evolutionary theory" (Laubichler and Maienschein 2009, 123).

Rather than a clearly defined research agenda, they refer to evo-devo as a "synthetic paradigm." On closer inspection, the research programs within evo-devo are too diverse to be subsumed under a single unified research program that would encompass all programs across the field. Within the synthetic paradigm, some biologists focus more on genetics, while others emphasize evolutionary issues. But in any case, all authors who claim to belong to evo-devo seek to establish relationships between these two components according to the different kinds of links or "unions" that may exist between development and genetics, development and evolution, or even between development, genetics, and evolution. In this, the program is faithful to Gavin de Beer's wish, expressed in 1930 book *Embryology and Evolution*, that biology should unify the different subdisciplines of embryology, genetics, and evolution.

All this shows how the "field" of evo-devo can be characterized as unified. However, it remains to be shown how the different explanations within the field can actually be integrated, since we have already shown (chap. 8) that integration may be better than unification.

De Beer used the concept of "plasticity" as a key concept for integrating the subdisciplines because of its independent use in the different subdisciplines. For similar reasons, we argue that the question of whether a true integration of developmental, genetic, and evolutionary explanations within EDS is possible through a precise analysis of the concept of "plasticity" by its theorists. Indeed, the concept of plasticity has been described in EDS as a "transversal concept" to the different subdisciplines of the field.[5] The problem is to know whether the term is used in a *specific way* (for a particular explanation or another), or whether it is used in a *general way* ("just" to underline a will to integrate the different types of explanation within evo-devo). It seems that many evo-devo biologists today continue to use the term *plasticity* ambiguously, using several definitions and uses of the same term, sometimes even in the same article. This becomes a problem precisely because of the increasing role that some theoretical programs in the field want to assign to the phenomenon of plasticity. Let us examine some of them, whose purpose was to offer a new synthesis.

Plasticity in Developmental System Theory

In 2003 the philosophers of biology Susan Oyama, Russell Gray, and Paul Griffiths edited a collective work titled *Cycles of Contingency: Developmental Systems and Evolution*. In this book, the editors proposed a new theoretical framework that they called developmental systems theory (often referred to as DST). The main aim of this new theoretical framework was to emphasize the central role of development in evolutionary processes. In developmental systems, each phenotype is conceived as the result of the interaction of many developmental resources, and the distinction between genetic and environmental factors is seen as one possible distinction among others, but not as the only crucial distinction of developmental resources (Oyama, Gray, and Griffiths 2003, 2). A system is defined as the sum of resources and processes that produce a life cycle (214).

When the book was first published, it provoked many reactions and even some controversy in the community of philosophers of biology, particularly with regard to the advanced objectives of the theoretical program. Developmental systems theory was clearly an approach that rejected a number of the theoretical assumptions of modern synthesis. The authors believed that modern synthesis was responsible for a "distortion" in the interpretation of many biological processes (see chap. 4). They believed that this distortion had somehow marginalized a number of central biological phenomena that needed to be prioritized in order to get a more accurate picture of biological processes. Phenotypic plasticity was one of them. Behind this ideal image, in which DST would appear as a relatively neutral or pluralistic description of biological phenomena in contrast to the biased one proposed by modern synthesis, one can in fact see the expression of a strong theoretical program (probably even more exclusive than that proposed by modern synthesis). Indeed, like most theoretical programs, DST adopted precise theoretical orientations and relied on specific disciplinary traditions that it wished to perpetuate (which will be examined in more detail below). Therefore, it seems necessary to return to the foundations of DST in order to better understand some of its assumptions (which are not always made), including those related to the phenomenon of plasticity. I think this clarification is important for the role that plasticity is often given in EDS in general.

Since DST (which claims to be a theoretical strand within evo-devo) aims to bridge development and evolution, an analysis of DST's claims might help to illuminate why plasticity is considered a key phenomenon in evo-devo in general. The first consideration in this regard is that developmental psychobiology has strongly influenced DST. Daniel Lehrman (1919–1972), to whom the authors of DST refer as a mentor, was an American ornithologist and comparative psychologist. In Lehrman's biography, the National Academy of Sciences (founded in the United States in 1863) states that "[Lehrman] influenced a whole generation of students in animal behavior in this country and abroad" (Rosenblatt 1995). The authors of the DST, therefore, in carrying on this particular tradition, depart from current approaches to embryology, which see the developmental process as limited to a particular period in the life of the organism. They seem to introduce the following theoretical assumption: the developmental process would not be limited to the early embryonic stages (or to the developmental process before birth) but would continue throughout the life of organisms (as evidenced by the development of their behaviors). In DST development is often seen as a general life cycle process; all living things develop, and development is not limited to immature bodies. Therefore, DST, which considers development to be analyzed together with evolution, does not really see the two phenomena as distinct (as I will show later), unlike mainstream evo-devo biology, which continues to maintain a clear distinction between the two processes.

How, then, do the DST proponents manage to give a precise definition of plasticity in such a context when it already seems difficult within the evo-devo programs, where the distinction between evolution and development remains? Because they believe that all living things develop, all living things change or vary; it is therefore no longer possible to distinguish a "difference" or a "change" caused by a developmental process from the "difference" caused by variance. In this context, it is no longer possible to distinguish *regulation* from *induction* or from *plasticity* (see chaps. 6 and 7). Therefore, the multiple implicit uses of the term *plasticity* in evo-devo, described as problematic above, can be understood in DST as a symptom of the collapse of distinctions between development and variation.

We are therefore left with the same kind of definitional problem that I described when Scott Gilbert (2006) described tertiary induction as a possible mode of embryonic induction during the developmental process (see chap. 6,

"Embryonic Induction Explains Variation in Developmental Process"). According to Gilbert's definition, tertiary induction could occur simultaneously with secondary and primary induction. I have shown that by accepting such a definition, it is easy to lose sight of the specificity of the scientific term *embryonic induction*, understood as "the interaction [via a mediator] at close range between two or more cells or tissues of different history and properties" (Gilbert 2006, 139), since tertiary induction does not imply an interaction between two cells/tissues but only a signaling of the environment to the cell(s) of the living bodies. Gilbert thus proposes a very broad definition of plasticity that fully characterizes all phenomena of variation: *development* becomes a synonym for *variation*, which in turn becomes a synonym for *plasticity*, a method also used by the proponents of DST.

Moreover, in contrast to dichotomies between genes and environment, or "organic" and "cultural," DST proponents see "both development and evolution as processes of *construction* and *reconstruction* in which heterogeneous resources are contingently but more or less reliably reassembled for each life cycle" (Oyama, Gray, and Griffiths 2003, 1). This "constructivist" framework depends primarily on what Oyama has called "causal democracy" (Oyama 2000a) and what Griffiths and Knight have called "the causal parity thesis" (Griffiths and Knight 1998), with the idea that genes are only one resource available to the developmental process. For this reason, they see a fundamental symmetry between the role of genes and that of the maternal cytoplasm. Developmental resources are a complex system (Griffiths and Gray 1994, 277).

One of the main consequences of the causal parity thesis is that a precise analysis of the only available biological characteristics (genes), understood as the cause of our traits, loses its importance. Philosophers Luc Faucher, Pierre Poirier, and Jean Lachapelle summarized the consequences of this approach in 2006. For them, DST rejects a form of modern preformationism that sees the information that programs development as "in" the genes. Instead, the DST thesis proposes to view development as a construction whose outcome (the phenotype) results from interactions between the organism and its environment as well as from the interactions within the organism (Faucher, Poirier, Lachapelle 2006, 162).

From the DST perspective, it is not necessary to consider genes as one cause among others, because ultimately any specific idea of determinism is false. Moreover, it is not enough to say that phenotypes are the *joint* product of genes and environment (Oyama claimed in 2000 that this was already widely accepted), but it remains to examine the relative influence of each of these causes over time. Thus, what was transmitted from one generation to the next was not the traits but what Oyama calls the "developmental means (or resources, or interactants)." This includes "genes, the cellular machinery necessary for their functioning, and the larger developmental context, which may include a maternal reproductive system, parental care," and so forth (Oyama 2000b, 29).

In this context, genes are no longer seen as the only units of inheritance, and selection pressures appear to act "on the whole developmental manifold at all levels of complexity." Evolution now appears as the "change in the life-cycles of organisms in their co-constructed niches, reflected by differential reproduction and distribution of developmental systems" (Robert, Hall, and Olson 2001, 956). In this highly original perspective, it is not necessary for the authors to refer explicitly to the concept of plasticity, since the idea of plasticity is implicit in this complex view of development and evolution.

Let us return to the original problem, which concerned the relationship between the idea of plasticity as a central phenomenon of evo-devo and DST assumptions about the role of development in evolutionary theory. Proponents of DST seem to think that an understanding of development precedes any understanding of evolution, or, in other words, that some empirical basis for evolutionary theory is still lacking (see Oyama [1985] 2000, 180). For them, understanding evolution is as important as understanding development, since it is the understanding of the developmental system that prevails. According to this position, in order to understand and explain any biological phenomenon, one must first make a precise study of the developmental process, the ontogeny, that underlies that phenomenon—an idea very similar to what Haeckel argued a century and a half ago.

However, as the philosopher of biology Jason Scott Robert and the editors Brian Hall and Wendy Olson rightly point out in *Keywords and Concepts in Evolutionary Developmental Biology*, although "mainstream evo-devo" can be linked to DST, the two approaches differ in many ways (starting with the respective roles they assign to genes in their conception of development). Although DST offers an original theoretical current that differs from the usual trends in biology, it has raised major empirical difficulties. For example, it has been criticized for not being able to propose specific evo-devo research agendas[6] (Schaffner 1998). Even more than the lack of concrete research agendas, one of the weaknesses of DST lies in its lack of solid definitions to characterize some of the biological phenomena it seeks to emphasize.

Proponents of DST have often claimed that, by emphasizing concepts such as phenotypic plasticity, nongenetic inheritance, or niche construction (among others), they offer a reconceptualization of development, heredity, and evolution (Oyama, Gray, and Griffiths 2003, 183). Too often, however, they seem to forget that some of these phenomena had already been described in the past, and that they had sometimes been described in a context where genes were considered a central factor, a key reference for analysis. This was particularly the case, as I have shown, for phenotypic plasticity (see chaps. 3 and 4).

Nevertheless, in some recent studies, the issue of defining (or of *redefining*) these biological phenomena within the new theoretical "developmental framework" has come to the fore among some evo-devo biologists (e.g., the discussion of "niche construction" in chap. 6). As far as phenotypic plasticity is concerned, Mary Jane West-Eberhard's work on "developmental plasticity" in an example of such a process of redefinition in a new theoretical context. The question, then, is to what extent the definition of plasticity that she offers also provides a clear reconceptualization of development within an evolutionary framework and/or what the consequences of such a redefinition would be for understanding the link between developmental and evolutionary explanations.

Mary Jane West-Eberhard's Theory of Developmental Plasticity

Mary Jane West-Eberhard was one of the first biologists to try to reconcile the biological processes of morphological plasticity with those of behavioral plasticity (West-Eberhard 1989). For many years, the American biologist studied the social behavior of bees in relation to their morphological diversity (size, presence or absence of reproductive organs and of secondary sex organs, etc.). In 2003 she published a book titled *Developmental Plasticity and Evolution*, in which she proposed a general explanatory framework that tried to take into account the interaction of organisms with their environment. The framework relied heavily on the notion of developmental plasticity. Over a period of almost twenty years, she studied the relationship between the behavior of insects (mainly bees) and their development, and she came to the firm conclusion that a new synthesis of evolution had become necessary that took development more into account. Such a view explains why her book is regarded as belonging to the field of evo-devo. However, West-Eberhard's ambitions go far beyond the presumed ambitions of evo-devo, for she not only brings together data from both disciplines—developmental biology and evolutionary biology—but also aims to propose a new evolutionary synthesis that includes development in its definition. For this reason, her approach is very similar to that of DST proponents. Indeed, in the opening pages of her six-hundred-plus-page book, West-Eberhard explains that in order to provide a synthetic theory of evolution, she had to offer an "inclusive definition of plasticity" in order to avoid "unnecessary distinctions at every turn," even if she assumed that such distinctions might be crucial at certain points (West-Eberhard 2003, 33).

But what exactly does West-Eberhard mean by the idea of an inclusive definition of plasticity? Does such a definition allow for a theoretical integration of developmental explanations within an evolutionary framework? These are the two main questions that I will try to answer.

Her book caused quite a stir in the biological community and among philosophers of biology when it was published and circulated (Pfennig 2004; Rollo 2004 de Jong 2005), because West-Eberhard's attempt was to propose a process for the emergence of evolutionary novelty that was no longer based on individual genetic mutations. In this new framework, genes are seen as followers rather than leaders in evolution (West-Eberhard 2003, 20). Changes in gene frequencies are seen as occurring subsequently, rather than initially explaining the emergence of a new phenotypic trait. While most theories explaining phenotypic diversity continue to rely primarily on quantitative genetic studies, West-Eberhard reverses the priority and puts genome evolution in the background to focus primarily on phenotype. Referring to the now classic definition of "phenotype" as "all traits of an organism other than its genome" (Johannsen 1911), West-Eberhard offered a theoretical framework in which the external environment and the genome are "on equal footing as architects of the internally orchestrated aspects of the phenotype, for all gene expression depends on conditions and materials of external environmental origin within cells" (West-Eberhard 2003, 32). The phenotype is conceived as a consequence of the expression of these factors, and its implementation depends largely on the developmental process. For this reason West-Eberhard considers it more relevant to start from phenotypic variation rather than genetic variation to explain the evolution of traits. She reminds us that selection on genes is indirect because it is the phenotype that is selected for first. However, her view leaves considerable room for nongenetic factors to explain the origin of the phenotypic variation that is then selected. A famous example she uses to illustrate her argument is the case of dwarf elephants that populated the Mediterranean islands in prehistoric times (described by Roth 1993). The origin of the new trait (reduced size) must be linked to a specific event, a famine that led to the progressive atrophy of the elephants, while the origin of the phenotypic trait "dwarfism" is natural selection favoring small elephants that are easier to feed (Roth 1993). Instead of trying to simplify the explanatory framework (by limiting it to a single cause, which would be the gene), West-Eberhard nuances it by integrating all the different elements at the origin of variation (environment, genes, development, etc.).

The general process is that a change in the environment of individuals leads to a "developmental plastic response"—"when some new input affects a preexisting responsive phenotype"—and to "phenotypic accommodation"—"the immediate adjustment to a change, due to the multidimensional adaptive flexibility of the phenotype" without genetic change—which, as a second step, enables the improvement of the individual's adaptation to the new environment (West-Eberhard 2003, 140). The recurrence of the initiating factor (whether environmental or mutational) produces a subpopulation of individuals that express the phenotype. The phenotypes resulting from this developmental plasticity are then selected. This results in a change in allele frequency, a "genetic accommodation," that improves and integrates the change.

West-Eberhard reports the famous example of such phenotypic accommodation observed by the Dutch veterinarian E. J. Slijper in 1942. He found a goat that was born with only two legs. This goat responded to its handicap with morphological and behavioral specializations that eventually allowed it not only to survive but also to move. Throughout the book, West-Eberhard returns to this fascinating case, which illustrates the general scope of her research. The essential point, which is the strength of this particular example, is that despite its severe birth defects, the goat's locomotor function is preserved during ontogeny. According to West-Eberhard, "adaptive phenotypic accommodations" that promote the normal development of a function (e.g., locomotion) are very likely to contribute to the emergence of a functional novelty (e.g., a new mode of locomotion). If viable and compatible with the ontogeny of the individual, this may lead to an increase in the frequency of the trait (e.g., two legs instead of four), which, after many generations, may lead to the production of a subpopulation of individuals with the trait in question. This process may be followed by "genetic accommodation," which results in a change in gene frequency that may affect the regulation, shape, or side effect of the new trait under the influence of the selection-variation process described by Darwin. In West-Eberhard's original process, the environment becomes a key player in the generation and selection of adaptations. The first step in the process is what West-Eberhard calls developmental plasticity.

In response to this approach, which is initially and in some later ways Lamarckian, the question arises as to what it means for West-Eberhard to regard plasticity as a key concept that requires an "inclusive definition." Two different interpretations can be suggested. Either it means for West-Eberhard that (1) the definition of the concept of plasticity should take into account the different traditions of its use in different contexts and in different subdisciplines of biology, or it means that (2) the definition of the concept of plasticity should encompass the different meanings that the term *plasticity* can take in the life sciences. Looking more closely at each of these two options makes it possible to clarify the different concrete situations in which biologists may need to see plasticity as a central concept, especially in a context where they are trying to link development with evolution.

Let me first explain what I mean by these two interpretations, both of which are related to the idea of an inclusive definition of plasticity (inclusive *uses* or inclusive *meanings*). Regarding the first interpretation, it seems that the term *developmental plasticity*, as used by West-Eberhard, has already appeared in the past in biology and especially in neurobiology (Bennett et al. 1964; Baudry, Thompson, and Davis, 1993; Foehring and Lorenzon 1999). It refers to "changes in neural connections during development as a result of environmental interactions as well as neural changes induced by learning" (Norman et al. 2007). The term helps to highlight the specific changes in neurons and synaptic connections as a result of developmental processes. This is one use of the term that is confined to the field of neurobiology. However, the term *developmental plasticity* has also been used in developmental biology (including genetic studies) to refer to developmental pathways or patterns (see chap. 6). It thus takes into account the complexity of genotype-phenotype interactions during development and is based on environmental conditions (Gilbert and Epel 2009). The second use/definition differs from that in neurobiology mainly because it refers to different entities (neuronal connections in neurobiology on the one hand and interactions between genetic and environmental factors in developmental biology on the other). Therefore, when it is used in different fields, it is associated with different assumptions and results. An inclusive definition of plasticity across these different subdisciplines would ignore the specific context and the specific subdisciplines that may be involved.

In terms of a definition that encompasses different meanings, the concept of developmental plasticity can be understood in two different ways based on the etymology of the term. In ancient Greek, it can be related to *plastikos*, meaning "related to shaping" or "for shaping," and to *plastikê*, from the verb *plassein*, meaning "to mold" or "to form." As I have shown, in a biological context the term can describe the body's ability to build its shape over time (plasticity as form generating), as illustrated, for example, in studies of limb growth in development (e.g., Forgacs and Newman 2005).

On the other hand, it can refer to the ability of a single genotype to have alternative possible phenotypes as a result of environmental signals (plasticity as form taking), as illustrated by the different patterns on butterfly wings depending on the season in which they develop (e.g., Brakefield et al. 1996).

However, notwithstanding the fact that West-Eberhard's conception of development appears to be in consistent with DST, a major difference is that she chooses to focus on phenotypes rather than developmental systems. Indeed, the main problem with DST lies in its focus on developmental systems. While at a theoretical level the description of the developmental system" seems satisfactory, from a strictly empirical point of view it remains difficult to identify the limits of such a system and therefore to study it in detail. It is probably for this reason that West-Eberhard focuses on phenotypes and not on the dynamic developmental systems. In addition, most contemporary biologists have been trained with the conceptual paradigm that emphasizes the genotype-phenotype distinction. It has therefore become extremely difficult for any contemporary biologist to think outside this paradigm. However, West-Eberhard's decision to retain Johannsen's definition of the phenotype is not without consequences for the formulation of her theoretical goals.

But before considering such implications, let me first consider what West-Eberhard means by the word *phenotype*. When one speaks of phenotype, one usually thinks of morphological phenotypes, that is, the structures of organisms that can be described in terms of some salient features (e.g., size, shape, color on certain structures, etc.). In practice, the reference to phenotypic plasticity involves the study of morphological plasticity, which allows for an account of morphological alternatives (e.g., polymorphism in the case of discrete phenotypes, or norm of reaction in the case of continuous phenotypes).

West-Eberhard extends the field of study of phenotypic plasticity, as used in genetics, by referring to the notion of developmental plasticity. She believes that the term *plasticity* should no longer be used in a very specific sense (specific to some subdisciplines such as neurobiology or developmental biology) but in a more inclusive and general way, leaving aside the differences between the different meanings that the term can have depending on the context and the subdisciplines of use. West-Eberhard therefore redefines plasticity in the following way:

> *Plasticity* (responsiveness, flexibility) is the ability of an organism to react to an internal or external environmental input with a change in form, state, movement, or rate of activity. It may or may not be adaptive (a consequence of previous selection). Plasticity is sometimes defined as the ability of a phenotype associated with a single genotype to produce more than one continuously or discontinuously variable alternative form of morphology, physiology, and/or behavior in different environmental circumstances (Stearns 1989). It refers to all sorts of environmentally induced phenotypic variation (Stearns 1989). Plasticity includes responses that are reversible and irreversible, adaptive and nonadaptive, active and passive, and continuously variable. Many authors

have proposed terms that distinguish among these variations on the theme of plasticity. . . . But I will adhere to the simple, inclusive definition given here. On can be trapped by specialized terms into having to make unnecessary distinctions at every turn, as important as they may be for certain points. I use the word *lability* only to describe facility for *evolutionary* change in phenotype, as in "phenotypic plasticity correlates with evolutionary lability." (West-Eberhard 2003, 33)

The definition does not focus on any particular level of the organism and thus allows West-Eberhard to include under the same label different types of plasticity—such as morphological, physiological, phenotypic, and behavioral plasticity—but also adaptive plasticity—which is linked to the genotype-phenotype map—and nonadaptive plasticity—which is independent of the genotype-phenotype map—and finally, possible and varied synonyms such as *flexibility*, *malleability*, and *deformability*. Such an inclusive definition of plasticity is possible because she also defined development broadly. As in the case of DST, it adds objects of study other than those traditionally assigned to the field of developmental biology. Where developmental biologists traditionally focus on "morphological phenotypes," West-Eberhard decides to add other types of phenotypes to her plate. For instance, she believes that "phenotypic structure" refers not only to specific morphological structures, but also to "the organization of the phenotype at any level of analysis, not just to an item of morphology" (West-Eberhard 2003, 31). Thus, behaviors and physiological processes can also be considered as phenotypic structures. The reference to the notion of "structure" clearly reflects the strong influence of behavioral ecology on her work. However, if West-Eberhard gives broad (and also quite modern) definitions of *development* and *plasticity*, her definition of *phenotype* remains quite conventional despite her willingness to apply it to phenotypes other than morphological phenotypes. Indeed, she refers to Johannsen's 1911 article in which (as shown in chap. 3) the definition of *phenotype* relies heavily on a precise understanding and definition of the genotype—the phenotype is somehow *definitionally determined* by the genotype. It therefore fits into the traditional genetic view of the term. While West-Eberhard follows Johannsen's definition of the phenotype, she continues, probably against her own will, to give considerable weight to the genotype-phenotype distinction (a distinction that DST advocates have de-emphasized). What are the consequences of such a definition of the phenotype for the inclusive definition of plasticity that West-Eberhard adopts?

Either West-Eberhard's inclusive definition means that different *uses* of the term are united under a single banner, or it imposes a synoptic view in which the different *meanings* of the term (sometimes used within the same subdiscipline) remain distinct from each other and retain their relative importance while being integrated into an inclusive definition.

Regarding the first option, the biologist and philosopher of biology Massimo Pigliucci, in his book *Phenotypic Plasticity: Beyond Nature and Nurture*, points out that there are different definitions or different "kinds" of plasticity (Pigliucci 2001, chap. 2). He then suggests that it is possible to "build among" these definitions by observing and comparing the phenomena described (Pigliucci 2001, chap. 11), because in all cases, whether it is an alternative choice of morphological structure, behavior, or of a physiological response, what is being described corresponds to phenomena that allow organisms to change in response to environmental challenges. Such a view implies that it is possible to refer to a single definition of *plasticity* that accounts for all these different uses, since the phenomena described are quite the same, or at least can be understood in the same way (for a similar view see also DeWitt and Scheiner 2004). It seems that West-Eberhard's inclusive definition of plasticity is not so new, and her approach is quite similar to Pigliucci's idea. What she adds in the end is essentially a broader definition of development that takes into account phenotypes other than those usually described.

Accordingly, a conclusion can be drawn: West-Eberhard's inclusive definition of plasticity can be used to assess the different uses of the term in different subdisciplines and contexts. However, such a definition does not say more about the ability of the concept of plasticity (understood as a unified concept) to integrate developmental and evolutionist explanations. Furthermore, West-Eberhard's definition remains consistent with the definitions of plasticity of the genetic tradition and thus does not make a complete break with modern synthesis. Finally, for her, an "inclusive definition of plasticity" means nothing more than shifting the focus from genes to phenotypes, which seems problematic at worst—with regard to Johannsen's definition of *phenotype*, which has been based on the definition of *genotype* and not on that of *development*—and at best insufficient to provide an operational concept that would help to think theoretically about a new synthesis of evolution that would fully include development.

In conclusion, West-Eberhard's definition cannot really be called "inclusive" in the sense of providing a synoptic account of the different meanings of plasticity that would be needed to integrate evolutionary and developmental biology. It still focuses mainly on the "passive" meaning of the concept rather than its "active" meaning, understood as the ability to generate a new form (as used in stem cell studies). However, she emphasizes the developmental component of the process involved in plasticity and places some emphasis on developmental plasticity in her conception of evolution to show how development produces an adapted phenotype, which is necessarily linked to evolutionary ecology. So far, West-Eberhard (among others) has shown that developmental plasticity exists and plays a role in biology. However, to truly link the two disciplines and show that developmental plasticity is a major factor in evolution, developmental plasticity would have to be shown to be primarily adaptive and to precede genetic adaptation in time. This remains a missing piece of the puzzle for most biologists and one that few biologists are willing to embrace as a research agenda (probably because they see it as too similar to Lamarckism). So far, studies have shown that developmental plasticity may exist and may have a genetic basis, but nothing more.

Reflecting this theoretical gap, West-Eberhard's definition of *plasticity* does not seem to include the "active" meaning of plasticity associated with the "embryological tradition," which focuses on developmental processes. A few years ago, the cell biologist Stuart Newman and his colleagues defined plasticity as "the array of pattern forming mechanisms that operate during the development of complex organisms" (e.g., adhesion, lateral inhibition, cohesion, etc.) (Newman, Bhat, and Mezemtseva et al. 2009, 553). Such a definition of *plasticity* can be seen as part of the "embryological tradition" of plasticity, particularly because it focuses on *processes* rather than entities. In this view, the concept of plasticity takes on an "active" meaning. Therefore, West-Eberhard's inclusive definition of *plasticity*, as inclusive as it may be, does not take into account the "active" meaning of the term, especially when it is associated with the idea of "developmental process."

My initial question was whether West-Eberhard's *inclusive* definition of *plasticity* shows how plasticity was a central concept in the context of coordination between developmental and evolutionary explanations. In other words, when the author mentions the idea of an inclusive definition of plasticity, does she have in mind the formulation of an operative concept of plasticity, or does she consider (as do the proponents of the DST) that one should only define plasticity broadly: as a synonym for *development* or *variation*? In the light of my analysis, it seems that West-Eberhard's assumptions about the role of plasticity, like those of the DST proponents, are based on the same subdisciplinary foci. Indeed, her training is largely in animal ethology and insect social behavior. Whatever her claims, West-Eberhard tends to define the concept of plasticity broadly in order to account for *specific* developmental processes (in particular, the emergence of certain animal behaviors, such as locomotion in the two-legged goat) that would have been neglected in the past. However, unlike the proponents of DST, her definition is compatible with empirical application (particularly in relation to molecular biology studies).

Finally, West-Eberhard offers an inclusive definition of *plasticity* that reflects her disciplinary influences. The consequence is that by referring to a broad definition of *development*, it becomes difficult for her to specify what she means empirically by *developmental plasticity*—namely, how the process described by the term differs from the developmental process. Moreover, it must be admitted that her concept of plasticity is ultimately not so different from that of her predecessors. While West-Eberhard (unlike most DST theorists) refers to concrete empirical data to support her theoretical view, she does not add anything to the definition of *plasticity* by building on evo-devo studies. As shown above, most evo-devo studies, including West-Eberhard's, address the issue of plasticity based on concepts already articulated in the tradition of their subdiscipline—developmental biology or evolution (the main reference to this issue concerns the genetic notion of "phenotypic plasticity")—rather than providing an operational and inclusive definition of *plasticity* that would be new and original. However, the obstacles to this definition are essentially related to the lack of comprehensive studies on the active meaning of the term *developmental plasticity* and not to the lack of a genuine desire on the part of biologists to give an account of the phenomenon of plasticity in biology.

Beyond West-Eberhard's original work, it seems that the proposal for "theoretical recontextualization" made by DST proponents has had some success among evo-devo biologists (notwithstanding the criticisms addressed to DST). Some of them have worked toward analyzing phenomena that DST theorists themselves had intended to introduce into biological research. To take different examples, the biologists Odling-Smee, Laland, and Feldman have addressed the issue of niche construction (Odling-Smee, Laland, and Feldman [2003] 2013), Eva Jablonka has become interested in the issue of nongenetic inheritance (Jablonka, Lamb, and Zeligowski [2006] 2014), while West-Eberhard has focused on the issue of developmental plasticity (West-Eberhard 2003). Gerhard Schlosser and Günter Wagner analyzed the phenomenon of modularity (Schlosser and Wagner 2004), while Hallgrimsson and Hall dealt with the concept of variation (Hallgrimsson and Hall 2005).

It is also worth noting that, among these evo-devo biologists, many express a growing interest in ecological developmental biology (also called eco-devo or eco-evo-devo—each hyphen reflecting the common will to include or to incorporate the different elements separated by the hyphen), a field dedicated to the analysis of biotic and abiotic environmental inductions as well as the environmental control of development (Van der Weele 1999; Hall [1992] 1999; Gilbert 2001). It is among these biologists that DST finds its most enthusiastic audience and supporters (Robert, Hall, and Olson 2001). It is highly doubtful that the theoretical approach proposed by DST was the catalyst for these biologists. However, it is fair to say that such an approach can help to unite a growing number of biologists around a common research program.

Within the broader framework of evo-devo biology, it seems that what biologists lack most, even more than an awareness of the need to recontextualize evolution in relation to development (as claimed by DST proponents), is a clear redefinition of the concepts traditionally associated with the phenomena they wish to observe and explain within this new recontextualization. However, I argue here that as soon as biologists began recontextualizing by focusing on phenomena neglected by modern synthesis (i.e., phenotypic plasticity, epigenetic inheritance or niche construction, etc.), they lacked accurate and appropriate definitions that would allow them to refer unambiguously to these phenomena within a developmental context that remains largely undefined. Furthermore, it is now clear that not all evo-devo biologists seem to agree on what they mean by "development."

This issue of theoretical redefinition is somewhat reminiscent of Johannsen's redefinition of *inheritance* after the rediscovery of Mendel's laws in 1900. In 1909 Johannsen began to argue that a conceptual distinction between genotype and phenotype was a priority in light of his own research on the factors of Mendelian inheritance and in order to improve our understanding of living phenomena:

> It is a well-established fact that language is not only our servant, when we wish to express—or even to conceal—our thoughts, but that it may also be our master, overpowering us by means of the notions attached to the current words. This fact is the reason why it is desirable to create a new terminology in all cases where new or revised conceptions are being developed. Old terms are mostly compromised by their application in antiquated or erroneous theories and systems, from which they carry splinters of inadequate ideas, not always harmless to the developing insight. (Johannsen 1911, 132)

While evo-devo biologists may have the same problem of conceptual redefinition, they do not always express it explicitly. Nor does DST manage to overcome the expectations of contemporary biologists.

In the following section, I will examine whether or not other theoretical research programs in the field manage to fulfill these expectations.

Plasticity in the Extended Synthesis of Evolution

In 2008 a group of sixteen evolutionary biologists and philosophers of science met at the Konrad Lorenz Institute in Altenberg, Austria, to reassess the status of the evolutionary process in contemporary biology, particularly in light of the empirical and conceptual breakthroughs in the field brought about the emergence of the field of evo-devo. The group called for the development of an Extended Synthesis of Evolution. This extension, they argued, should be linked to the results of the revolution wrought by the emergence of molecular biology and to the vast body of empirical knowledge on genetic variation in natural populations, phenotypic plasticity, phylogeny, and developmental biology.

Both the authors of the Extended Synthesis and the proponents of DST are at odds with the framework of modern synthesis because they share the view that the focus on genes has obscured how a precise analysis of the developmental process could help to increase our knowledge of evolutionary mechanisms. In addition, more than sixty years after Julian Huxley published his book *Evolution: The Modern Synthesis*, the authors of Extended Synthesis had another reason to meet. They express it in the following quote:

> Many of the empirical findings and ideas discussed in this volume are simply too recent and distinct from the framework of the MS [modern synthesis] to be reasonably attributed to it without falling into blatant anachronism. Concepts such as evolvability . . . for instance, did not exist in the literature before the early 1990s; phenotypic plasticity . . . was known, but consistently rejected as a source of nuisance, not of significant micro- and macro-evolutionary change. (Pigliucci and Müller 2010, 4).

The argument they develop here is slightly different from the one put forward by proponents of DST. Concepts such as phenotypic plasticity alone cannot account for significant processes in a developmental perspective on evolution. Rather, they argue that new concepts have emerged and should be defined or redefined within the new conceptual framework that includes development. Although some authors refer to the concepts of "developmental plasticity" or "phenotypic plasticity" as new, the authors of the extended synthesis mostly undermine the novelty of the concepts. See, for example, Pigliucci (2001), who underlines that the concept was already present in the past. In short (and despite the fine nuances they bring to bear in each case), they believe that the concepts of phenotypic plasticity and developmental plasticity are the sorts of concepts that are simply too new to be grasped in the context of modern synthesis. Therefore, in their view, it has become necessary to develop a new conceptual framework.

However, if such concepts are "too recent" (as the authors claim) to be assigned to an old conceptual framework—that of modern synthesis—it becomes difficult to consider whether their *current* associated definitions should be assigned to a new conceptual framework. At most, the conceptual framework would need to be updated. For example, when Johannsen made the distinction between the genotype and the phenotype at the beginning of the twentieth century, he based the distinction on a new conceptual scheme in which he reconceptualized heredity in relation to the emergence of Mendelian genetics. The definition of *genotype* could therefore only follow (and not precede) the elaboration of the new conceptual framework. Moreover, it was only later, in the 1930s and 1940s, that the genotype-phenotype distinction helped to bring together the subdisciplines of genetics and evolution and to expand our understanding of evolutionary mechanisms. It seems that new conceptual schemes can only precede, rather than follow, the formulation of NEW concepts that can then lead to new types of explanation. The two moments—the *formulation of a new conceptual framework* within which a number of new concepts are formulated, and *the new types of explanation* that follow—should be clearly distinguished. The authors of the extended synthesis do not seem to make this distinction clearly. Indeed, in the preface to their book, *Evolution: The Extended Synthesis* (Pigliucci and Müller 2010), the editors claim that "by incorporating the new results and insights into our understanding of evolution, the explanatory power of evolutionary theory is greatly expanded within biology and beyond" (viii).

In this sentence, there is clearly a gap between the first ambitions of theoretical reconceptualization and reformulation and the subsequent subsidiary and elevated ambitions of extending the old theoretical framework. Thus, the authors intend both to break away from the old framework and to extend it with new concepts (which, however, cannot be interpreted within the old conceptual framework). The solution for the authors seems to be to adopt concepts whose definitions remain consistent with the old framework but that also allow them to offer new types of explanations for evolutionary mechanisms. Another possibility would have been precisely to distinguish each new concept, which would have helped to highlight the new conceptual framework in relation to the old one (as Johannsen, for instance, did in his time).

Let us now look at how the authors refer to the concept of plasticity. Chapter 11 of *Evolution: The Extended Synthesis* (Pigliucci and Müller 2010) is devoted to "phenotypic plasticity." The first question that should be asked is, Is this an old or a new concept? The editors of the book seemed to hesitate on this question in their introduction, describing the phenomenon as new but the concept as old (4). The second question is, Should the term *phenotypic plasticity* (most commonly used by biologists of all disciplines and "traditionally" associated with population genetics; see chap. 3) be preferred to the less well-known term *developmental plasticity*? Again, no precise answer is given, since Massimo Pigliucci, the author of the chapter, has chosen to title it "Phenotypic Plasticity." In the first sentence he says that the concept is old, but in the conclusion he emphasizes that the phenomenon has strong implications for extended synthesis; he also explains that biologists have largely misunderstood the concept, even though he claims at the end of the first paragraph, "In fact, ever since Woltereck coined the term 'reaction norm,' it should have been clear that plasticity is a property of a genotype, and that it is specific to particular traits within a given range of environments" (355).

Despite the ambiguity surrounding the concept of plasticity, Pigliucci endeavors to show the importance that the phenomenon of phenotypic plasticity can play in our understanding of evolution. In doing so, he focuses mainly on the expression of the phenomenon of plasticity rather than on defining the concept of plasticity itself. Having somehow resolved the apparent confusion surrounding the concept, considering that "it should have been clear that plasticity is a property of a genotype" (355), it remained for Pigliucci to show why the phenomenon is important for an "extended" understanding of evolution. The author intends to show that by identifying the different "mechanics of plasticity," one could then understand "how deep a role phenotypic plasticity plays in the restructuring of evolutionary theory" (362). Finally, he intends to show that it is possible to identify a certain unity behind the different mechanisms, which depends on the role that all these phenomena have for an extended understanding of evolution.

It seems that even if many different possible mechanics of plasticity can be identified, as Pigliucci shows, and different biological phenomena can be associated with these mechanics, which could ultimately have significant evolutionary implications, no new definition of the concept is proposed. Moreover, in such a view, the sense of continuity between modern synthesis and extended synthesis does not allow us to see the reasons why a conceptual extension would be necessary, since the original framework itself was formulated precisely to accommodate further developments. Julian Huxley himself, in the various introductions to the successive editions of his book *Evolution: The Modern Synthesis*, has indicated what these extensions were. The consequence of emphasizing the role of plasticity in this theoretical framework is that confusion about the term increases rather than diminishes and that old definitions, rather than being accepted, are changed or distorted to the point of possibly losing all precise meaning.

Conclusion: Plasticity Disunified

At the beginning of the chapter, I wondered whether the term *plasticity* had a specific use in the field of evo-devo (i.e., whether it was associated with one type of explanation or another), or whether it was used in general (i.e., whether it potentially helped to highlight the integration between levels of explanation within the field). At the end of this chapter, I can now confidently say that there is no single conception of plasticity in evo-devo and that differences in how biologists understand plasticity often reflect not so much differences between theoretical programs but rather different uses within each of the subdisciplines that have now been brought together under a common banner (consistent with the differences that exist between subdisciplines in biology in general). Nevertheless, in analyzing the different theoretical programs within the field, I have found that the phenomenon of plasticity has attracted the attention of theorists who have made strenuous attempts, without apparent success, to redefine the concept. While there is no reason to speak of a definitive failure in this regard, it is clear that none of the proposed definitions has yet been able to dispel the sense of confusion associated with the use of the concept.

This observation leads me to believe that the characterization of plasticity as a "boundary concept" that I proposed in a previous chapter (chap. 4) is justified. The concept of plasticity demarcates many different domains within contemporary biology: it not only defines the boundaries between different subdisciplines (developmental biology, evolution, genetics) but also demarcates different types of explanation (explanation of developmental or physiological variation versus explanation by natural selection); finally, it demarcates "nature" from "nurture." All these boundaries are also areas of tension and confusion rather than areas of integration despite what the authors of EDS or DST claim. The best we can say is that the concept of plasticity, like the idea of frontier in geography, can serve as a link between different biological territories. It can reveal the fact that the different territories, even if they are distinct, are not completely independent of one another, and even that they often interact with one another, but always, I insist, at the risk of further confusion. The boundary thus acts as a signal, allowing the evolutionary biologist to remember that knowledge of gene frequencies is not enough to explain the generation of a trait and the developmental biologist to look beyond the singularity of the development of a trait to understand the process in a general way. But the "hyphen" that plasticity represents (like the hyphen between evo and devo in evo-devo) does not ultimately allow for integration, and there is no possible synthesis at this particular boundary between different domains (at least not at the current state of knowledge). If it is possible to "think" the articulation of these different domains by referring to the concept of plasticity, there is no single phenomenon of plasticity to account for these different domains considered together. The use of the term *plasticity* in EDS remains more a consequence of biologists' will to think the articulation of development and evolution than a solution or an operational concept that would effectively account for a new form of integration between different explanations in biology.

Plasticity in the Life Sciences
Resistance or Recurrence

Why not simply say what one means and leave it? So then he tried saying the grass is green and the sky is blue and so to propitiate the austere spirit of poetry whom still, though at a great distance, he could not help reverencing. "The sky is blue," he said, "the grass is green." Looking up, he saw that, on the contrary, the sky is like the veils which a thousand Madonnas have let fall from their hair; and the grass fleets and darkens like a flight of girls fleeing the embraces of hairy satyrs from enchanted woods. "Upon my word," he said (for he had fallen into the bad habit of speaking aloud), "I don't see that one's more true than another. Both are utterly false." And he despaired of being able to solve the problem of what poetry is and what truth is and fell into a deep dejection.

VIRGINIA WOOLF (1928)

Let me try to put together the various pieces of the plasticity jigsaw puzzle that I have collected throughout this study to see whether, by putting them together, one can get a coherent and complete picture of what plasticity is in the life sciences. But is this even possible? Virginia Woolf, in her imaginary biographical novel *Orlando*, shows the confusion the poet finds in trying to describe the world around him, his fear of not being able to access the truth in its entirety. My aim is more modest than that of the poet; it is not to reach the whole truth, but to succeed in shedding new light on the concept of plasticity in the life sciences. To do this, I have chosen to examine the different uses of the concept of plasticity and the articulation of these different uses through the recent history of biology.

Why the reference to Virginia Woolf's *Orlando*? Apart from the anecdotal but timely fact that the main character, Orlando, changes sex and becomes a woman in the middle of the story (this is reminiscent of the plasticity of the clown fish that I mentioned in chap. 5), the precedent quote refers to the difficulty that the poet or writer (but this remains largely valid for the man of science) may encounter in trying to give an exhaustive and satisfactory account of the reality of the world. Realizing the insurmountable difficulty of the task, the character falls into a deep depression. However, biologists can sometimes experience a similar feeling when confronted with the complexity of life. In their analysis of biological reality, they try to describe and explain all the phenomena of life as precisely as possible. But, it seems, they are constantly and inevitably forced to refer back to the idea of plasticity, as if only this idea would allow them to overcome the difficulty they have in explaining biological reality in its entirety.

Thus, in the fourth century BC, Aristotle—the great philosopher but also, in a certain sense, the father of biology—obviously had neither the methods of analysis nor the scientific concepts that we have today. However, he already referred to the concept of plasticity (even if it was borrowed from the field of art) in order to explain the particularity of living beings to form, deform, or take on different shapes. Contemporary biologists—who now have a wide range of analytical tools, models and concepts for observing and describing living things—continue to use this concept of plasticity in a not so different sense, although they give it a less intuitive definition. How can we interpret the fact that a term whose use originally seemed to reflect a certain theoretical aporia, itself linked to a scattered and fragmented knowledge of the living world, is still commonly used today by scientists when they try to describe or explain living phenomena? Should we, like Orlando, who noted the difficulty of describing the world once and for all, fall into a deep despondency when we realize that contemporary biologists continue to struggle with the same fundamental difficulties that Aristotle encountered in his time? Or should we see in this persistent use of the same term throughout the ages as a reassuring sign, proving the heuristic importance of the concept of plasticity, whose use remains constant in the biological literature? It seems that this observation, instead of plunging us into a deep confusion, can, on the contrary, be a trigger to generate new perspectives of reflection.

The constant presence of the concept of plasticity in the biological literature, as far as one can see, can be interpreted in two different ways. The first interpretation implies that there is a kind of resistance of plasticity in biology, as one would speak of the "resistance of the wind," which is the means by which the wind manifests its own presence. In this interpretation the persistence of the concept of plasticity is a sign of the *resistance* of a theoretical approach that biologists have tried to maintain throughout the history of the discipline despite the constant changes that exist in the understanding of biological reality. The second interpretation that I propose is to see this constant presence as the *recurrence* of the same concept throughout history and in the different sciences of life: the concept of plasticity can then be seen as an index of an invariable and indispensable constant to account for the specificity of life, even if it remains difficult to define clearly.

Finally, the problem is to understand what epistemic status should be given to the concept of plasticity in contemporary biology. For this purpose, I distinguish two main types of current issues related to the use of the concept of plasticity in the life sciences: the metaphysical issues and the epistemological issues.

Metaphysical Issues Raised by Plasticity

In the first chapter I referred to the metaphysical implications of the concept of plasticity: I highlighted the different conceptions or theories of plasticity that were associated with a certain intuition that the first theorists of living beings might have had about plasticity. I showed the emergence and the influences of these different intuitive conceptions of plasticity and the way in which they were able to impregnate natural philosophy and then natural science.

THE DUAL NATURE OF THE CONCEPT
OF PLASTICITY

The first important observation I could draw from this historical analysis is that the concept of plasticity necessarily implies an ontological duality. In other words, by its very definition, plasticity offers a dialectical perspective. For in the image of the ancient uses of the term, based on its Greek etymology, two different meanings remain invariably associated with the concept of plasticity in the living world: an *active meaning* that speaks of a capacity, a disposition, that the living being possesses to deform itself, and a *passive meaning* that speaks of the fact that living beings take on various forms. These two senses are the two sides of the same coin, that is to say, they form a pair from which thought, by comparing them, will be able to elaborate "a feeling of the whole" or a unified intuition. The reference to the concept of plasticity in the field of biology will thus make it possible to give an account of one sense or the other (and often even of both in an ambiguous way), but the intrinsic (or ontological) duality of the concept will also make it possible for the scientist to refer to it in order to symbolize a possible synthesis between the two senses of the concept.

However, I have seen that the concept of plasticity is associated with numerous analytical approaches in biology that seek to account for the intuition that biologists have of a possible reconciliation (this time) between the two notions of an opposition (ultimately accounting for what we might call the intuition of a synthesis in the making).

To take an example, the perspective that seeks to bring development and evolution together in the same explanatory scheme by relying on the concept of plasticity is based on the intuition that the reference to the concept of plasticity would make it possible to take into account in a synthetic way the "active" part (that of the developmental process, which refers to a potentiality) and the "passive" part (that of the evolutionary point of view, which takes into account diversity) of the description of ontogeny and variation of form. By referring to the concept of plasticity in order to explain the intuition of a synthesis in the making, biologists are not only asked to consider development and evolution together (which they never manage to do without confusion) but also to consider how the synthesis thus achieved can allow us to go beyond. In other words, the biologist who talks about plasticity has to understand and even solve problems that could not be solved within each of the disciplines taken separately. Although such an approach seems promising at first sight, it is not without major difficulties, both scientific and philosophical.

Could the concept of plasticity be understood not as a concept that dialectically unites contradictory properties but rather as an expression of an intuition of a metaphysical order specific to living beings throughout the history of biology? It would then be better understood as a "simple concept," as Bergson describes it in *La pensée et le mouvant* (*The Creative Mind: An Introduction to Metaphysics*). Bergson pointed out the problem that arises when one tries to grasp an intuition of a metaphysical order by means of concepts: the problem, in his view, is that a concept that is too simple may not require any effort for a metaphysical investigation of the object in what it has that is essential and proper (Bergson [1938] 1998, 186–87).

Is the concept of plasticity one of those "too simple" concepts of which Bergson speaks here? Certainly it can be considered as a convenient way for the biologist to *symbolize* the intuition that she can realize an essential synthesis between the active and the passive part of the generation of the variation of the form, or, in a more general and contemporary way, between development and evolution. In this sense, the reference to the concept of plasticity offers the biologist, at low cost, the possibility of *solving* the problem she is facing: namely, the precise identification of the theoretical pairs (or oppositions) that she is trying to articulate in a synthesis. Let me explain. What if the biologist today had an *intuition*, in the Bergsonian sense, of a true unity that exists between development and evolution and not just a vague idea of what the rapprochement between development and evolution should be? The difficulty will then be to know what she is aiming at when she speaks of plasticity (a specific phenomenon? a set of explanations relating to different fields of knowledge? or a unified theoretical principle?). Let us consider, for example, the cases in which the biologist refers to the concept of plasticity in order to explain both the active and the passive part of the generation of variation in form. If she tries to report on this without referring to the concept of plasticity, she will have to be able to identify what she means by the "active part" and the "passive part" of the generation of variation in form. For the question of the "generation of variation in form" can itself be interpreted in different ways depending on whether one takes an ontogenetic, phylogenetic, or even phenogenetic perspective. Thus, from an ontogenetic perspective, understanding the generation of variation of the form will require the biologist to have a prior understanding of the *developmental origin of variation* (which means in the active sense, the developmental process that leads to change, and in the passive sense, the different patterns that can be adopted). From a phylogenetic perspective, understanding the generation of variation of the form will require the biologist to understand the *evolutionary origin of variation* (which means in the active sense, the processes of mutation, recombination, or changes in genetic frequencies, and in the passive sense, the differences between species, clades, families, etc.). Finally, from a phenogenetic perspective, understanding the generation of the variation of the form will lead the biologist to understand the *origin of a given phenotypic variation* (meaning in the active sense, norms of reaction, and in the passive sense, differences within a species for a given trait).

However, if the biologist chooses to use the concept of plasticity, he no longer feels the need to specify exactly which type of pair (onto-, pheno-, or phylo-, etc.) he is referring to, and which would be the active or passive part in each case—difficult tasks. By short-circuiting these problems, so to speak, by referring to the apparently simpler concept of plasticity, the biologist finds a certain advantage. For where the philosophers of nature were first forced to identify one by one the different concepts of the specific metaphysical dualities of living beings—for example, between the active and passive parts of the generation of form—the biologist can now refer to a simple concept (the concept of plasticity) that gives him the reassuring illusion that it gives a unified and synthetic account of two a priori antinomic positions.

Let me give you a very concrete example: when the biologist refers to the concept of "phenotypic plasticity." The concept allows us to use a single term to describe two different aspects of the same process: the ability *to generate a diversity of phenotypic forms*, and the ability *to account for the diversity of phenotypes*. But the definition of the term usually proposed, as "the capacity of a given genotype to generate more than one possible phenotype," goes beyond the simple intuition one has of plasticity. It adds or implies an important element coming from a theoretical debate other than that of the opposition between the active and passive meaning of generation, an element that concerns the *causal determination of the phenotype*, which is based on the idea of genotypic information. If the concept of plasticity in general has allowed the biologist to put an end to the idea of a possible and even necessary synthesis between two different ways of understanding the generation of variation in form in the life sciences, the specific definition of the concept of "phenotypic plasticity" refers to a more specific intuition. It does not simply give a name to the idea of a possible synthetic conception between the active and passive parts in the description of the generation of variation in form but gives a name to the idea of a synthesis between the active and passive parts of phenogenesis in which the active part of the process is described with reference to a genotypic cause.

Therefore, the concept of phenotypic plasticity seems to correspond well to these too simple concepts described by Bergson. Bergson underlined that simple concepts divide the concrete unity of the object, which may mean, for our purposes, that the reference to the concept of phenotypic plasticity will never allow the biologist to give an *absolute* account of the intuition he has about the generation of phenotypic variation. But Bergson also pointed out that simple concepts (and the concept of plasticity can be considered as such) "also divide philosophy into different schools, each of which keeps its place, chooses its chips, and begins with the others a game that will never end" (Bergson [1938] 1998, 188). Now, one of the contemporary metaphysical problems with the use of the concept of plasticity in biology is that it seems to force philosophers to reformulate old philosophical problems from the contemporary analysis of phenomena they have. For example, where previously the debate seemed to pit partisans between different *modes of generation* (epigenesis versus preformation), the reference to the concept of plasticity allows contemporary biologists to insist on an opposition that no longer concerns the mode of individual generation but the *ultimate causes of the generation of variation in form*.

This first observation leads us to identify a second category of metaphysical issues. The previous category corresponded to the *way* in which the concept of plasticity can be conceived from an ontological point of view: either in its active sense or in its passive sense. In both cases we were interested in the general meanings of the concept—the concept of plasticity is characterized by a certain duality carried by its etymology and its intuitive understanding. The metaphysical implications of using such a concept seem to be that its use in biology leads to the emergence of new debates rather than the resolution of old ones. I will now examine the metaphysical issues[1] that depend not on the general meaning attached to the term *plasticity*, but on its intuitive meaning in the specific context of biology.

Plasticity as Deformation

The question I am now asking is not what kind of phenomenon they are try-ing to describe with the concept of plasticity but *how* biologists think about plasticity in the life sciences. In other words, it is no longer a question of considering plasticity as a "simple concept" backed up by the singular intu-ition of a possible synthesis but of understanding *what* intuition of plasticity biologists have and *what terms* they can use when thinking about it. In the study of life, plasticity generally refers to a capacity for *deformation*, that is, a disposition for a given living object to lose its initial form or to change its expected form. This intuition, as I shall show, is the source of some important metaphysical questions today.

In contemporary biology there is no longer any question, as there was in the sixteenth century with William Harvey, for example, of a "plastic force" responsible for the organization of living matter. Nevertheless, as in Har-vey's time, or even two centuries later in Caspar Wolff's, biologists continue to raise questions about the *origin*, the *seat*, and the *manifestation* of this property of "deformation" (while talking about plasticity). First of all, I have shown in the contemporary biological literature (chap. 5) that two opposing ideas clash over the question of the origin of the property of deformation. Some biologists believe that plasticity is determined by genetic causes, while others believe that plasticity is an emergent property, or a "by-product" of evolution, and therefore not directly caused by genes. The main problem at the root of this opposition lies in the difficulty biologists have in distinguish-ing the property of deformation as such from the result of deformation. In fact, when biologists stopped referring to the idea of a "plastic force," they also stopped referring to either physical or biological (vitalist or mysterious) forces, implying that the plastic property could be explained from an exclu-sively biological point of view.

Second, if we consider the question of the seat of deformation and ask where contemporary biologists locate plasticity, there are still other com-parisons to be made. In the seventeenth century, the Cambridge Platonists saw plasticity mainly as an intellective property, a power of the mind.[2] The thinking of these philosophers certainly has little to do with the basis of an understanding of the plastic property at the level of the brain. The property of plasticity in living beings is far from being summarized by what some re-searchers call neuronal plasticity[3] and therefore in the brain. The synonymy is only circumstantial: "plasticity" refers to very different problems in neurology and in the philosophy of biology, although there are some similarities. With the emergence and development of genetics at the beginning of the twentieth century, a number of new and increasingly specialized disciplines in biology have flourished. This property of living organisms, plasticity, is thus observed at all scales of life, at the level of the organism considered in a general way but also locally at the level of cell layers, tissues, cells, genes themselves (with particular reference to gene regulation), gene networks, populations (and in, generally, all dynamic systems), phenotypes, even behaviors, and so forth. In short, there does not seem to be a single level in the organizational hierarchy of living organisms where plasticity has not been ascribed. The question, then, is whether the *deformation* that biologists characterize as plasticity, observable at a given scale and defined at that same scale, is the expression of a single property of the living being that could be manifested at different levels of organization, or whether it is a matter of different properties of deformation. In other words, are we talking about the same thing when we talk about plasticity at one scale of life or another? Are we not often talking about very different biological phenomena that are lumped together under the same generic term? The use of the concept of plasticity in contemporary biology also overlaps in this sense with the old problem of reductionism in biology and in particular with all its possible variants or alternatives, which in contemporary biology are no longer limited to the simple opposition between reductionism and antireductionism (long characterized by vitalism, now by biological normativism or holism) (see Brigandt and Love 2008).

Bergson, speaking of intuition, pointed out that abstract ideas (which here would correspond to each of the individual properties of deformation at different levels of organization) could certainly be useful for analysis. And indeed, in the case of plasticity in biology, it seems that the identification of different properties (or different types) of deformation at each scale of organization of living things has allowed biologists to better understand the question of variation of form and organization at each of its different levels—studies of phenotypic, individual, or cellular plasticity are examples, among others. But abstract ideas (singular "types" of deformation) could not replace intuition, the metaphysical study of the object in what is essential and proper. Therefore, this analysis invites us to consider that, as far as plasticity is concerned, all the studies of the different types of plasticity or of the different properties of deformation will not allow us to reach the knowledge of plasticity in what it has of more "essential and proper." In other words, it is not because we have a cartography of the concept that we have a perfect definition of it. It is dangerous because the concept generalizes and abstracts. Therefore, it always distorts it more or less by the extension it gives it. The various concepts that we form of the properties of a thing thus draw around it many wider circles, none of which applies exactly to it (Bergson [1998] 1938, 187–88).

In fact, the analysis of the different "types" (or "properties") of deformation in living organisms seems, paradoxically, to lead to a considerable loss of precision with respect to the general object (plasticity in the life sciences) that biologists are trying to understand. However, this precision, which is essential for scientific work, seems to be precisely what is lacking when it comes to plasticity in biology. Thus, in order to gain access (as far as possible) to the "integral experience" (Bergson [1998] 1938, 227) of plasticity, it seems necessary to go beyond the methodical description of the various properties or mechanisms of deformation that exist in nature. Therefore, clarifying what *distinguishes* these different types of deformation expression from each other rather than what *brings them together* (which is what most biologists and theorists have focused on so far) may be a good direction to take toward a more precise and integrated understanding of plasticity in the life sciences.[4]

A third point is to consider the question of the *manifestation* of deformation itself, and relatedly, the question of the precise circumstances under which deformation can manifest itself. Intuitively, plasticity can be thought of as a dispositional property, meaning that the expression (or manifestation) of the property depends on the presence of certain conditions. To put it in a more formal way, and drawing on the definition of disposition proposed by Carnap, "F expresses a disposition if there are an associated manifestation and conditions of manifestation such that, necessarily, an object is F only if the object would produce the manifestation if it were in the conditions of manifestation" (Carnap 1936).

Emphasizing the dispositional nature of plasticity leads us to focus not only on the manifestation of deformation but also on the *conditions* under which deformation manifests. The question will be whether emphasizing the manifestation of deformation and the conditions under which deformation manifests allows us to account for some very common and general features of plasticity.

We often observe that in biology, when it comes to plasticity, the conditions for the manifestation of the property, as well as the nature of the associated manifestation, are usually given at the beginning. In other words, the property of plasticity is most often understood as a disposition. Some titles of scientific articles testify to the precision with which the conditions of manifestation and the manifestation of the deformation or modification associated with the dispositional property of plasticity are specified. Here are three examples. First, "Predator-Induced Shell Dimorphism in the Acorn Barnacle *Chthamalus anisopoma*" (Lively 1986). In this first example, the property of plasticity is manifested by a phenotypic dimorphism of the shell of the barnacle (two different forms can be adopted), and the conditions of manifestation of one or the other form depend on the presence or absence of a predator in the environment. Second, "Cell Surface Mechanics and the Control of Cell Shape, Tissue Patterns and Morphogenesis" (Lecuit and Lenne 2007). In this second example, the property of plasticity is manifested by a change in cell shape that depends on specific molecular signaling at the cell surface. Third, "Physiological Adaptation and Plasticity to Water Stress of Coastal and Desert Populations of *Heliotropium curassavicum* L." (Roy and Mooney 1982). In this third example, the property of plasticity is manifested by a physiological adaptation of the plant, and the conditions of manifestation depend on the growth context of the plant, which provides more or less important water resources. These three examples are particularly interesting to show how the analysis of the *conditions of manifestation* of deformation and the analysis of the *manifestation* of deformation do not make it possible to highlight a unique and unified property of plasticity. Therefore, it seems that we will not be more successful in identifying plasticity in living organisms as a univocal phenomenon by simply characterizing plasticity as a dispositional property, or, in other words, by focusing on the manifestation of deformation and the conditions under which deformation occurs.

Nevertheless, such an analysis sheds new light on the question of dispositional properties in biology. Let us take another look at the conditions of manifestation in each of the three examples cited above, which are fairly representative of the many studies that exist on the phenomenon of plasticity. In the first two articles, the authors emphasize the importance of the *stimulus* (predators in the first example, surface cell signaling in the second). The third article focuses more on the *context*.

This leads to an initial observation: it will be difficult to say in general terms whether the condition for the manifestation of the property of plasticity in living organisms is based more on the existence of a *given context* or on the mobilization of a *defined stimulus*. The nuance seems subtle, but the questions it raises are important. In the first example, the authors seem to insist on the importance of the stimulus provided by the presence of the predator in the barnacle's environment. However, it is quite conceivable that barnacles in a different environment from the one in which they are normally found (e.g., if they are stuck on rocks that are more calcareous than their usual rocks) but exposed to the same stimuli (the same predators) will not necessarily express the property of plasticity, because the calcareous rocks will have a repulsive effect on the predators, and thus the signal that induces plasticity (the predators) will no longer act. In other words, the risk of defining plasticity in terms of a given stimulus is the risk of isolating in the abstract a phenomenon that only takes on its full meaning in the constitutive relationship of the organism to its environment.

In the second example, however, the stimulus for cell surface signaling seems to be more intrinsic to the organism's milieu because it depends not on the environment but on the constitution of the cell. At first sight, this example of plasticity might seem to lead to a better understanding of the dynamic relationship between the organism and its adaptation to the environment. However, here, too, we must ask ourselves which context (which type of organism) favors or does not favor deformation. In other words, will we necessarily obtain the same result in the case of a unicellular organism as in the case of a multicellular organism, for example? This question is not resolved and is barely addressed in the article.

The last example (the plant that adapts according to the presence of water in its environment) is more complex because here we find both the context and the stimulus fixed. As a result, it is difficult to determine which stimulus is responsible for the manifestation of the deformation: the presence of water alone (considered in an abstract way) or the presence of water *combined with a certain climate*? This indeterminacy in the conditions of manifestation of plasticity has a cost: it is always difficult to understand, from one article to another, what one is entitled to conclude, not only about the phenomenon of plasticity in general but even about the phenomenon of plasticity for all the individuals of the species considered.

Following these different examples, we may wonder about the specificity of the dispositional property of plasticity in living organisms. Should we take into account the context, the stimulus, or both to account for this disposition? As with the dispositional property of "being poisonous," which has been the subject of much debate among philosophers of science (especially concerning the importance to be given to the context in relation to the stimulus), one might again wonder whether considering plasticity as a disposition would not be more of a statement of the problem rather than a real solution (see Prior 1985; Choi 2005; 2011; Lewis 1997).

It should be added that, in biology, too, it is possible to distinguish situations in which the biologist is led to explain the conditions of manifestation of the property of plasticity from situations in which he is not necessarily led to do so. This is reminiscent of certain distinctions generally made by philosophers of science, such as that between *canonical* and *conventional* dispositional properties. The three examples above clearly illustrate situations in which plasticity can be regarded as a canonical dispositional property provided that the conditions of manifestation and the manifestation of the property are specified.

It should be noted, however, that in the contemporary literature on plasticity there are also examples of studies that deal with this property as such without mentioning the conditions under which it is manifested. Let us take two representative examples: a 2002 article entitled "Developmental Plasticity in Plants: Implications of Non-cognitive Behavior" (Novoplansky 2002), and another, also from 2002, titled "Consequences of the Inherent Developmental Plasticity of Organ and Tissue Relations" (Sachs 2002). A simple comparison of the titles of these two articles with the three previous examples is sufficient to show how the analysis of the conditions of manifestation or the manifestation of the disposition is a different problem from the analysis of the disposition itself. Indeed, in the previous cases, the study of the manifestations and conditions of manifestation of the dispositional property of plasticity did not allow us to say that their identification improves our understanding of plasticity as a conventional dispositional property (i.e., of the property of plasticity insofar as its conditions of manifestation are not specified).

On the other hand, it may be possible to shed light on the conventional dispositional property of plasticity in the life sciences by proposing specific definitions for the canonical dispositional properties of plasticity (i.e., for properties of plasticity that would be identifiable by given stimuli or contexts). To date, existing analyses of the canonical dispositional properties of plasticity seem to have little in common. However, if one were to focus on the specific determination in each situation of what the source of the stimulus is and what constitutes the context, it would be possible to isolate certain studies in which entities of the living world that manifest a certain deformation (modification) when exposed to given stimuli are considered plastic, regardless of the context. Therefore, in this specific case, the identification of stimulus and context would make it possible to highlight some general characteristics shared by the conventional dispositional property of plasticity.

The concept of plasticity raises the same questions for contemporary biologists as it did in the past: what is the basis of plasticity? To which entity does it belong? How does it manifest itself? And it leads to the introduction of new metaphysical questions into contemporary biology. These all boil down to the expression of a single problem concerning plasticity: how can models of multiple phenomena be subsumed under a single concept? It seems that no one has yet found a solution to what appears to be a metaphysical aporia.

It remains for me to consider a final category of metaphysical issues. These are the questions that touch on the particular relationships, at the metaphysical level, that the notions of "plasticity" and "living beings" maintain.

Plasticity, the Specificity of Living Beings?

In the first part of this study, I pointed out that the notion of plasticity was not used, or was even deliberately ignored, by some embryologists from the

eighteenth century onward in order to immunize their work against any theoretical connotation (in particular vitalism) that was considered philosoph-
ically problematic. For example, the embryologist Caspàr Wolff deliberately
refused to refer to the notion of plasticity, whereas his predecessor William
Harvey, during the Renaissance, did not hesitate to make ample mention of
a *vis plastica* (a plastic force) when describing the processes at work during
development. In the eighteenth century (at the time of Caspar Wolff, to be
precise), the notion of *vis plastica* was brought closer to the notion of vital
force. It is then more related to questions about the definition of life than to
questions about the deformation of matter (the latter being what interested
Wolff in the first place). Thoughts on the definition of life were stimulated
in the eighteenth century by the school of physicians of Montpellier (and in
particular by Paul-Joseph Barthez in 1778). As Canguilhem has shown, the
physicians of this influential school sought to remove the underlying theological connotations from the account of life proposed by Stahl (Canguilhem
[1968] 1973). Thus, Barthez makes the vital principle a unique principle governing animal life, the true productive cause of the phenomena of life. In this
sense organic matter as such is left with few organizational capacities. If the
beginning of the eighteenth century was the high point of vitalism,[6] the nineteenth century was marked by the disappearance of the biological current of
vitalism, which quickly disappeared after its radicalization by Xavier Bichat.
Nevertheless, the survival of vitalism in certain works of embryology in the
nineteenth century (Driesch [1914] 2010[7]), but also and above all in philosophy in the twentieth century (Bergson 1969[8]), proves that *vitalist thought* has
never completely disappeared from reflections on biology. On the contrary,
it has always remained present in biology, even if at certain moments in its
history as a shameful part.

Since the second half of the twentieth century, *ontological vitalism* (the
assumption of the existence of forces or substances specific to living beings)
and *epistemological vitalism* (the idea that physical-chemical methods do not
allow us to explain the totality of vital phenomena, especially those marked
by a finality) have been universally rejected by biologists. On the other hand,
another form of vitalism (or rather a possible redefinition of the old vitalism), *methodological vitalism* (the idea that vital phenomena are the result of
the global organization of the living being, the study of which consequently
requires the establishment of its own methodology different from that which
has made the success of the physical sciences), has gradually gained a certain
place in specialized literature since the beginning of the twentieth century,
particularly studies on behavior. However, it is clear that in contemporary biology, the concept of plasticity is often associated (among other things) with

those specific works that deal with the great diversity of behaviors (whether they are manifested in plants or animals, or even humans) (e.g., Metlen, Aschehoug, and Callaway 2009; Dingemanse et al. 2010; Lerner 1984). As such, the concept of plasticity seems to have become an essential concept for explaining not the specificity of living beings in general but the specificity of behaviors or, more generally, of *vital phenomena*—the analysis of which, from a methodological point of view, does not require recourse to a mechanistic reductionism. Thus, in contemporary biology, where plasticity is discussed, whether it is the work of the proponents of DST or that of those who claim to follow the original approach initiated by the biologist Mary Jane West-Eberhard, they all aim to provide a complete and scientifically satisfactory explanation of the behavior of living beings.

We can see a certain kinship between this type of approach among contemporary biologists and theorists and the philosophical theories that have been developed since the seventeenth century, notably by the Cambridge Platonists. Indeed, for the Cambridge Platonists, thinking about the organization in the living world could not be based solely on the accumulation of knowledge about singular mechanisms. Sooner or later, reference had to be made to an integrating principle. The concept of plastic nature fulfilled this role. Now, the main difference between the approach of contemporary theorists and that of the Cambridge Platonists (apart, of course, from the minor detail that plasticity is no longer regarded as a subordinate agent of the divine will) is that the latter do not position themselves, like their predecessors, in the rejection of *any form of mechanism*. On the contrary, all the contemporary reflections on plasticity seek to give an account of an *integrating principle* by referring to a concept of plasticity that gives a large place to mechanisms while at the same time arranging a space for methodological vitalism.

Let's take the example suggested by David Sloan Wilson in *The Philosophy of Behavioral Biology* (Plaisance and Reydon 2012). In a chapter titled "Evolving the Future: Sketching a Science of Intentional Change," Wilson emphasizes that the phenomenon of phenotypic plasticity (which he defines as the general capacity of organisms to modify some of their characteristics in response to changes in the environment in which they live) is central when it comes to human behavior because the latter expresses a very high degree of plasticity compared to all other animal species. Wilson seems to echo Rousseau's distinction between the rigidity of instinct in animals (the pigeon, e.g., would die of hunger rather than change its diet) and the great adaptability of human behavior when it is determined by free agents. But the distinction is also reminiscent of the thought experiment carried out in his time by

Cudworth, who tried to apply the question of "mental causality" at work in plastic nature to productions of nature that were, according to him, "devoid" of intelligence, as in the case of animals, thus demonstrating the existence of a gradation between living beings and placing man at the very top of this scale. However, it is a very different argument that Wilson has in mind when he insists on the difference in plasticity that exists in human behavior as opposed to that of animals. On the one hand, he believes that many human behaviors are "rigidly flexible," that is, they have "built-in capacities" to produce different signals depending on the circumstances. On the other hand, he also believes that many human behaviors can be conceived of in the manner of a "Darwinian machine" (see Calvin 1987; Plotkin [1993] 1997) in the sense that they instantiate an evolutionary process that allows for *open adaptation* to environmental circumstances (the human immune system is considered to be a good example of a Darwinian machine in that it instantiates, in its very functioning, an evolutionary process based on variation and selection). But the really interesting point is this: Wilson interprets these two situations as phenotypic plasticity. He believes that it is these two types of phenotypic plasticity (the rigid flexibility of human behaviors and the Darwinian machines they represent), *combined* with the human ability to transmit behavioral changes to future generations through cultural heredity and cultural evolution, that make humans highly adaptable to changes in their environment. Thus, to the Cambridge Platonists' conception of plastic nature, centered on the idea of *integration*, was added the idea of *Darwinian evolution*, which is now constitutive of what some authors understand by plasticity and that, according to them, allows us to understand in depth the *specificity* of certain vital phenomena and above all the great variability of human behavior. It is no exaggeration to say that, in a very derivative sense, the concept of the "Darwinian machine" refers to the idea of a black box in the description of vital phenomena, an idea that was already present in the concept of "mental causality" of the Cambridge Platonists. In both cases, the concept of plasticity refers to a description of vital phenomena that rejects a purely mechanistic approach.

Let us summarize. In the three preceding sections, I have tried to distinguish between the *intuition that stimulates contemporary biologists to think about plasticity*, with the metaphysical question it raises, and *plasticity as a possible route to a synthetic conception of development and evolution, or as a black box for the nature of vital phenomena*. In both cases, it is important to note that the concept of plasticity appears as a heuristic concept for the biologist, and this on several levels: it brings light to new forms of questioning on essential problems of biology; it takes into account a property considered specific to biological phenomena; and it reflects on the dispositional property that characterizes the living as a whole.

Until recently, the various disciplines, which had become specialized and acquired a certain autonomy, offered biologists the possibility of formulating explanations of the phenomena of the living world. Recognizing the paradoxical heuristic advantages of such a "disunity" in biology, both for access to knowledge and for its own good functioning, some biologists and theorists have sought to regroup, classify, and synthesize data in order to go beyond mere analysis and attempt to achieve a more unified understanding of living beings. The emergence of this new ambition in biology has been observed in the development of theoretical programs that aim to bring development and evolution closer together (evo-devo) or establish more or less unifying theories (DST) or even propose a new integrating conceptual scheme (Extended Synthesis). Within these different approaches, the concept of plasticity has always seemed important, even essential. And, in a sense, the concept of plasticity also raises heuristic issues by providing an account of the articulation of different levels of explanation in biology. In this respect, one can speak of an "internalist" perspective on plasticity because it seems that the importance that the concept of plasticity has gained in contemporary biology is mainly based on the conviction (which has only become stronger among biologists since the mid-1960s) that the understanding of plasticity phenomena opens up the possibility of nuancing the genetic influence on the formation of phenotypic traits. This explains why all those who have argued that genocentrism should be rejected have in fact regarded plasticity as a central concept with high stakes. But if there is one lesson to be learned from this book, it is that if the concept of plasticity is useful in biology in an empirical and local way, it is always ambiguous when considered in an abstract and generic way.

This brings us to the more properly epistemological issues surrounding the concept of plasticity.

Epistemological Issues

THE CONCEPT OF PLASTICITY AS AN EPISTEMIC PARADIGM

The idea of the "epistemic framework" was developed by Jean Piaget in a book written with Rolando Garcia titled *Psychogenesis and the History of Science* (Piaget and Garcia 1983). To define this original idea of "epistemic framework," the authors rely on an analysis of the scientific revolution that took place in the seventeenth century. They show that the revolution that took place in science, and in mechanics in particular, "was not the result of the discovery of new answers to the classical questions about motion, but of the discovery of new questions that made it possible to formulate the problems in a different way." It goes without saying that this way of looking at the scientific revolution of the seventeenth century is very similar to the way Thomas Kuhn looks at the question of "scientific revolutions" in *The Structure of Scientific Revolutions* (Kuhn [1962] 2012). But Piaget and Garcia differ from their predecessor by focusing on the idea of a change in the "epistemic framework." Moreover, they do not forget to explain what distinguishes the epistemic framework from Kuhn's "paradigm." For the latter, each scientific epoch is characterized by a particular paradigm, that is, a certain conception that defines the theories, models, and concepts to be adopted for scientific research. The criteria that determine the direction of research are determined by the dominant paradigm of the time. The concept of epistemic framework includes, but is not limited to, that of the Kuhnian paradigm. Indeed, the authors consider that Kuhn's "paradigm" is mainly related to the sociology of knowledge, whereas the concept of "epistemic framework" allows to take into account (and also to distinguish from the social paradigm) a proper *epistemic* paradigm.[9] The "social paradigm" (which corresponds to Kuhn's paradigm as understood by Piaget and Garcia) is determined by factors that are exogenous or extracognitive, that is, of social or cultural origin, and that influence the direction in which scientific inquiry is conducted. The development of scientific knowledge is thus influenced by external demands imposed by society, which may lead to the study of certain types of phenomena or particular problems. These external demands will play an important role in the direction that the development of scientific theories takes. In this respect, knowledge will indeed be conditioned by this social paradigm, as Kuhn affirmed. Piaget and Garcia agree with Kuhn that this is not a one-way process: scientific theories can, in turn, influence the common patterns of thought in society.

Piaget and Garcia cite the example of the Aristotelians conception of the world, which, according to them, was completely static, since they believed that the "natural state" of the objects of the physical world was a state of rest. They compare their view of the world with that of the Chinese, for whom the world was in constant motion. Movement, the continuous flow, was the natural state of everything in the universe. These two different views of the world led to very different physical explanations. Piaget and Garcia point out that this "difference . . . does not lie in a methodological difference, nor in a difference in the conception of science. It is an *ideological* difference that translates into a different epistemic framework" (Piaget and Garcia 1983, 281, my translation). Thus, for a long time, the epistemic framework of the Greeks prevented Western scientists from thinking about the very idea of the principle of inertia.

The authors, of course, bring this idea closer to the "epistemological obstacle" à la Bachelard. But where Bachelard considered that there is a total break (*obstacle*) between prescientific and scientific conceptions (Bachelard [1938] 2002) and saw in prescientific irrationalism the essence of the "epistemological obstacle" to be removed (even if, for Bachelard, the progress toward greater rationality in the sciences is never fully achieved), Piaget and Garcia, on the other hand, consider that there is a relatively great continuity between prescientific and scientific thought—especially given that the mechanisms of the cognitive process are the same and that there is a certain type of "break" that is reproduced each time one passes from one state of knowledge to another in science. In other words, the "break" envisaged by Piaget and Garcia is smoother than that envisaged by Bachelard, since the epistemic framework constituted in their conception prevents the dissociation of the contribution coming from exogenous factors from that coming from endogenous factors. Piaget and Garcia point to the functioning of what Kuhn would call normal science[10] and what they call the epistemic framework: "Thus constituted, the epistemic framework begins to act as an ideology which conditions the further development of science. This ideology functions as an epistemological obstacle that does not allow any development outside the accepted conceptual framework" (Piaget and Garcia 1983, 282–83).

Although, as the authors point out, it is often difficult to make a concrete distinction between a "pure" ideology and an ideology of a political-social nature (since the influence of the political-social context is not always sufficiently clear to be able to isolate the various factors that have influenced the phenomena under study), they stress that in certain situations (as between the Aristotelians and the Chinese) this is nevertheless possible. Building on Piaget and Garcia's idea of an epistemic framework, I can highlight some of the reasons why biologists maintain the notion of phenotypic plasticity despite the many confusions that have been associated with this particular understanding of the concept of plasticity for several decades.

Indeed, the fact that biologists have continued to refer to the concept of phenotypic plasticity for the past sixty years is indicative of a form of "ideological resistance" of the concept of phenotypic plasticity in biology. It is a "pure" ideology because, despite the sociopolitical reasons that may have led to the formulation of the concept of phenotypic plasticity, these political divisions have now been overcome. Indeed, when Bradshaw formulated the concept of phenotypic plasticity in 1965, we were still in the midst of the Cold War between the United States and the Soviet Union (as this war did not officially end until the late1980s). In the last twenty years or so, many US (or, more generally, Anglo-Saxon) biologists and theorists have begun to take a new interest in the work of Soviet biologists, which had been carried out since the beginning of the twentieth century. In other words, if at the beginning of the twentieth century it was mainly sociopolitical reasons that were the reason for rejection of Soviet work by US scientists (which led them to adopt a very specific notion and definition of plasticity far removed from the theoretical connotations that underpinned the work of their Soviet counterparts), this political ideology, if still present, is less directly exerted on the representations of researchers. The difficulty that contemporary biologists, regardless of their culture and nationality,[11] have in conceiving of plasticity in terms other than the now common definition of *phenotypic plasticity*— which mainly takes into account the interaction between genotype and phenotype—is therefore, in my opinion, more related to the existence of an epistemic paradigm—generalized to the community of biologists publishing in international journals—than to a sociopolitical paradigm. The concept of phenotypic plasticity has become an operative object for the biologist, and it becomes all the more difficult to "rethink" it differently. It is therefore possible to consider that the concept of phenotypic plasticity corresponds to a "soft epistemological obstacle" (in the sense of Piaget), or what we call a "resistance" of the concept in biology, in the sense that it does not prevent the functioning of science as a whole but is today the source of confusion in the biological literature.

I emphasize that the concept that manifests this particular "ideological resistance" that I have described here is the one associated with the concept of phenotypic plasticity, but there may be other signs of resistance of old theoretical approaches in the contemporary literature (as, e.g., with the concept of morphological plasticity in the stem cell literature). In my opinion the general idea of plasticity in living organisms is quite different. Indeed, for the latter, the term heuristic recurrence seems better suited to clearly define the epistemic role that the notion of plasticity plays in the analysis of the living world. I must now explain what I mean by this.

Plasticity, a Heuristic Recurrence in the
Analysis of the Living World

At the end of their book, Piaget and Garcia insist on the need for "an epistemological reformulation" of the history of science. According to them, the visible continuity in science should not be considered as a postulate (as some of their epistemological colleagues seem to do[12]) but as the result of an "ex post facto investigation." Based on their comparative analysis of "psychogenesis" and the history of mathematics, the two authors assert that a certain continuity that can be observed in each of these fields "*does not exclude discontinuities* in the [described] processes" (Piaget and Garcia 1983, 292).

I can apply the same precise reasoning to the concept of plasticity in the analysis of the living world. Indeed, my research has allowed me to demonstrate that the use of the concept of plasticity has been recurrent in biology in the sense that its repeated use throughout the history of the description of the living world is not simply the result of repeated contingencies or linguistic fashions but rather the existence of unique connections between fields of study and traditions of use that I have methodically described and identified. Indeed, I have shown the existence of disciplinary, metaphysical, and ideological traditions of use across centuries and multiple fields of study within which one finds a certain unity in the meaning attributed to the concept of plasticity. The definition that accompanies a given occurrence of the term is not a simple definition by synonym(s) (e.g., by using the word *flexible* instead of *plastic*) but generally refers (more or less explicitly) to the *disciplinary tradition of use* to which the occurrence in question relates—and this can be seen in particular in the addition of adjectives to the noun *plasticity* (e.g., phenotypic plasticity, developmental plasticity, or adaptive plasticity, etc.) In other words, by referring to the idea of recurrence for plasticity, I insist on the continuity of a certain tradition of thought or epistemic framework that ultimately explains the continuing importance of plasticity in biology.

It is also possible to speak of recurrence in the use of the concept of plasticity in biology in the sense that when biologists use the term, they are giving an account of the same phenomena that have already been described in the past through the different fields of biology (e.g., the capacity of the living organism to modify itself during its development in embryology, the capacity to modify the phenotype of a given genotype according to the environment in genetics, the capacity of adaptive modification according to the external constraints in physiology). From then on, it is possible to speak of recurrence in the sense that recurrence makes it possible to explain the same phenomenon in a repetitive way.

This recurrence is thus a sign of a manifest continuity of the concept of plasticity throughout the history of biology. But it is not the continuity of a *singular use* of the concept of plasticity but rather the continuity of *multiple uses* of the concept of plasticity. It is precisely the diversity of the different uses of the concept of plasticity that makes it possible to take into account the specific connections that exist for each use, making these singular uses operational and heuristic units that fit into a particular epistemic framework.

Therefore, the main difficulty in identifying the existence of continuity between the different uses of plasticity lies in the fact that it is sometimes difficult to identify what distinguishes these different uses (what makes these singular uses operational units) given that the traditions that subsume them may overlap and intertwine with each other. Consequently, it is also sometimes difficult for biologists to have a clear and unambiguous idea of which tradition of use they are actually referring to when they refer to the concept of plasticity.

Finally, it should be noted that the general *continuity* of the use of the concept of plasticity in biology is also characterized by the presence of *discontinuities*, since, as I have shown, some uses tend to change over time. For example, the concept of phenotypic plasticity presents a difficulty (or an obstacle) for biologists who, for several decades now, have been trying to modify its definition by introducing a certain break with the tradition of the concept's use. But again, it seems that the break is not complete in the sense that it would introduce a "revolution" in the way biologists now understand plasticity. Rather, it seems that the new concept refers to the old one and even integrates it in some way. For example, when the biologist West-Eberhard defined the term *phenotypic plasticity* in an original way in order to find a new use for it, she had to refer to the previous definition of the term even though she was consciously *breaking* with the tradition of the use of the term. According to Piaget, such a mechanism guarantees the "continuity of knowledge": "In the case of the 'overcoming' [*dépassement*] of a structure of knowledge by a larger one, the overcoming [*dépassé*] is integrated into the overcoming [*dépassant*], which allows the continuity of knowledge" (Piaget and Garcia 1983, 304).

In this sense, it seems that the continuity of the concept of plasticity through biological analysis constitutes a heuristic element for the biologist who refers to it, since the reference to plasticity does not simply allow her to eliminate philosophical problems—the one concerning the ontological status of generation is a striking example (i.e., the distinction between epigenesis and preformation)—but also allows the biologist to inscribe her research in a certain continuity with her predecessors as well as with her contemporary colleagues in other fields of biology.[13] In this respect, I would like to point out that if the idea of continuity mainly evokes a persistence in time between the way biologists thought about plasticity in the past and the way they think about it today, it is also possible to consider that in biology it is necessary to take into account the continuity that must also exist between different fields of biology. It seems that this second type of continuity is often more problematic for biologists than the first, the continuity between domains. In fact contemporary biologists are now confronted with a new problem mainly related to the specialization of methodological and conceptual tools that has taken place in parallel with the specialization of subdomains. In such a context the recurrence of the concept of plasticity across the different domains will once again seem to be able to constitute a heuristic element for the biologist, since it will allow her to account for a certain unified "representation" of living phenomena across all the domains of biology.

What do I mean by this? It seems that the recurrence of the concept of plasticity in all the disciplines of biology is due to the fact that biologists consider plasticity to be a useful (better, necessary) image, to access the knowledge of the living world? As Canguilhem wrote, "life is the opposite of a relationship of indifference to the *milieu*." If there is a concept that allows us to characterize this type of relationship, which in biology is anything but a relationship of indifference to the environment, it is the concept of plasticity. Plasticity allows us to explain one of the essential characteristics of life. Now, if the phenomenon of plasticity makes it possible to explain a characteristic of life in its essential relationship with the milieu, does the concept of plasticity make it possible, more generally, to *recontextualize* knowledge in biology in relation to scientific knowledge in general? In other words, wouldn't the concept of plasticity allow the biologist or the philosopher to rethink the question of the relationship between the living and the nonliving?

In his article on "Life," Canguilhem underlines one of the reasons why access to knowledge of living phenomena is a great difficulty for the scientist. In fact the biologist-scientist who tends to carry out a scientific analysis of her object will do so by referring to tools and models that were initially formed on the basis of hypotheses that *denied* life (Canguilhem [1968] 1973). From then on, it is possible to consider that the biologist—who is to some extent aware of the differences between her models and those of the physical sciences (which nevertheless often guide her analysis)—likes to refer to the concept of plasticity because it introduces the possibility of *representing the living world independently of the laws of physics but without denying these same laws*. In this sense, the concept of plasticity makes it possible to establish a link between the purely physical and the biological. This is one of the first reasons to consider the continuous use of the concept of plasticity in different disciplines as what we have called a heuristic recurrence in biology.

If this first difficulty highlights the existence of potentially different *conceptual tools* for understanding inanimate and animate beings, Canguilhem in his article highlights a second difficulty that can be associated with all approaches to the analysis of living matter. This second difficulty allows us to consider the existence of technical obstacles to the understanding of living beings. In fact the first relationships that humans established with living beings were technical relationships: fishing, hunting, agriculture, and so forth. But the way in which inert objects and living objects are used differs in an important way. When humans want to use an inert object, they first isolate its properties (e.g., the hardness of stone for a weapon, the lightness of wood for a boat). They can cut the stone if it is too big, or sharpen it into the shape of an arrow, and they can remove the bark from the wood to make it smooth, so that the boat glides more easily on the water. On the other hand, in the case of the "use" of living beings, the organism must always be considered as a whole (and even in industrial breeding—at least until today—it is not possible, for example, to isolate the "egg-laying" property of the hen by cutting off its head so as not to have to feed it). Canguilhem summarized the consequences of this difficulty as follows: "It is hardly necessary, therefore, to insist on the force of the inclination that the use of life by the living human being has ingrained in him, by which any attempt at an analytical explanation of life is at first unconsciously censured." The reference to the concept of plasticity, however, allows us to overcome this difficulty, since this concept also refers (both literally and metaphorically) to human creation as well as to the creation of all living things. It allows the biologist to go beyond the censure of any analytical explanation of life by considering the living object not only in its decomposition (its analysis into elementary parts, its reduction to quasi-inert matter) but also in its construction (its synthesis, its own quality of organism). In this sense, the concept of plasticity makes it possible to establish an original link between "strictly material" matter and "living" matter.

From these two observations it can be said that the concept of plasticity, insofar as it recurs in the various disciplines of biology, is a heuristic concept for biology in the sense that it allows us to account for the relationship (i.e., both what unites and what distinguishes) between living and nonliving matter. Thus, just as François Jacob showed in *The Logic of Life* that the concept of reproduction is a means of uniting the "visible structure" with the "invisible" (Jacob [1981] 2022), I can defend the idea that the heuristic recurrence of the concept of plasticity in the study of the living is characterized by its capacity, unique in contemporary biology, to unite living and nonliving matter.

Conclusion: The Plastic Material as a Metaphor for Overcoming the Confusion

This chapter began out of a certain perplexity or concern because I noticed that the concept of plasticity, the use of which seemed to reflect a certain *theoretical aporia* in the past, continued to be used in a recurrent (and even flourishing) way in contemporary biological literature at the risk of becoming too often a source of confusion in texts written by biologists. The persistence of this concept of plasticity in the discourse on the living world led me to question its singular epistemic role.

At the end of this journey, I can somehow overcome my initial perplexity by looking at the concept of plasticity from a new angle. First of all, the analysis of the metaphysics of the concept of plasticity led me to link past and present conceptions of plasticity in the life sciences, thus identifying a number of constant heuristic challenges in the understanding of living things. The reference to the concept of plasticity seems to be a convenient way for biologists to evacuate old philosophical problems and formulate new ones related to the contemporary interpretation of the phenomenon of plasticity in the life sciences. Moreover, the concept of plasticity also has some crucial attributes for understanding the specificity of the analysis of living beings in relation to other objects of knowledge—in particular, the fact that plasticity helps to shed light on the specificity of biological dispositional properties (for which it is extremely delicate to distinguish the question of stimulus from that of context, the question of incidental factors from that of the environment). Finally, the use of the concept of plasticity in biology is a good tool for distinguishing the heuristic difference between ontological vitalism and methodological vitalism: a specific characterization of vital phenomena can be made without the need to define life or the mechanisms of life in an univocal way. By highlighting the similarities between the different conceptions of plasticity within the different fields of contemporary biology, I have shown that the integration of developmental and evolutionary explanations through the concept of plasticity should not be seen as the result of the integration of different biological factors but rather as a boundary concept, a place where these different factors manifest themselves distinctly from one another. From this observation, it is now possible to conclude that the concept of plasticity is a *useful* (or rather, *efficient*) concept in biology to take account of the complexity of living phenomena without necessarily adopting a reductionist approach but rather by considering a dynamic nesting of the different levels of explanation considered by biologists. Finally, perhaps an original way of thinking about the more diffuse uses of a general concept of plasticity (such as developmental plasticity), as opposed to the now common use of the concept of phenotypic plasticity (which has a positive connotation in biology), would be to use the metaphor of plastic material. This metaphor, applied to living organisms, would allow the property of plasticity to be interpreted as a *facsimile of naturalness*. This new metaphor may prove useful in the future to better capture plasticity in the life sciences. In fact, if the concept of phenotypic plasticity seems to depend on a precise epistemic framework at work in contemporary biology, which sometimes constitutes a certain obstacle for contemporary theorists, the generalist concept of plasticity—reinterpreted in the sense of a facsimile of *naturalness*—can be conceived as a heuristic concept of a new and original scope for contemporary biology insofar as it allows us to explain the relationship between the analysis of the living world and that of the nonliving world while escaping from a purely physicalist approach.

Acknowledgments

I would like to express my sincere gratitude to all my teachers, friends, colleagues, and relatives who have read all or part of this manuscript or with whom I have been able to discuss and exchange views on the subject. In particular, I would like to thank Jean Gayon and Philippe Huneman, who accompanied me throughout this research and encouraged me from the beginning to turn it into a book. The content of this manuscript has changed considerably since my initial research, but they have both had a profound influence on my approach to philosophy, which is as close as possible to its subject, biology, but also on the value of examining the genealogy of concepts, their history, and theoretical issues. Since the beginning of this work, Gregory Bateson, Werner Callebaut, Jean Gayon, and Michel Vervoort have died. They have all helped me to think through this research. I hope that this book will pay tribute to the intellectual exchange we were able to have.

More generally, I would like to thank the community of philosophers of biology, which for me is like a working family, for the kindness and collegiality of many of its members over the past few years.

The Institute for the History and Philosophy of Science and Technology (IHPST) in Paris was first a place of emulation, then a place of shared friendship, whose members are now my close friends and family. I would like to take this opportunity to thank all its members, those who have been part of the institute or who have visited it, especially between 2007 and 2014.

I am also grateful to the community of the International Society for the History, Philosophy, and Social Studies of Biology (ISHPSSB), which, from my first conference in Salt Lake City in 2011 to the most recent in Toronto in 2023, has always been a place of exchange, enriching encounters, and friendship. I have met colleagues there who have become lifelong friends. Among them is Denis Walsh, who welcomed me to the IHPST in Toronto on two occasions in 2010 and 2011, and thanks to whom my thoughts on plasticity were able to mature. He also introduced me to some of his fellow biologists, such as Ellen W. Larsen at the University of Toronto and Stuart Newman, who opened the doors of the New York Medical College to me and allowed me to observe the plasticity of the eggs he was working on in his laboratory. Stuart's perspective on development has had a profound influence on this work.

Finally, I would like to thank all the people who, in one way or another, helped me to produce this book: Labex "Who am I," which funded the work of an English-speaking proofreader, Mara Wood (I am extremely grateful for her patience and help in clarifying the issues I raise in this book); Karen Darling, academic editor at the University of Chicago Press, for her kindness and encouragement, as well as the editorial board and the entire editorial team; the three reviewers of earlier versions of this manuscript for the many hours they spent carefully reading the book and suggesting improvements that I hope will make it more enjoyable to read; and Juan Diego Soler and Jane Kassis, my close friends, who agreed to read the English versions of the submission forms at the last minute.

Last but not least, I would like to thank Steeves Demazeux, who, although already implicitly mentioned as a former member of the IHPST, has a very special place in these acknowledgments, as well as my parents, Jacqueline and Anton Nicoglou (the latter of whom made the cover photo and most of the illustrations in this book) and Hélène Bamberger, for their unfailing support in the progress of this work.

However, having said that, I take full responsibility for the shortcomings and errors that surely remain.

Notes

Chapter One

1. I call plastic- terms that have the root "plastic," such as "plasticity."
2. The "idea" here is seen as the outcome of the male's semen, as that which could provide the form.
3. The "matter" is specific to the female who was "in-formed" by the semen.
4. I call it "middle-passive" because there is no mention of the agent, but it could just as well be seen as a passive voice in relation to the topic at hand.
5. In his *History of Embryology*, J. Needham listed eleven major contributions of Aristotle's to embryology, such as his attempts at classification, his comparative studies between species, or the distinction between primary and secondary sexual characteristics, and so forth.
6. "Rational conception" as opposed to the pre-Socratic mythological conceptions.
7. The distinction and debate between generation by epigenesis and generation by preformation arose later in the eighteenth century with the early works in embryology.
8. "Cartesian mechanism" is sometimes also called "Cartesian materialism."
9. As I will show, in many respects More and Cudworth's positions were closer to Aristotle's view than to Plato's, but this is usually explained by the idea that Aristotle's philosophy was the continuity of Plato's.
10. Descartes referred to an idea, which existed from antiquity and had been revived in the Early Middle Ages, that assumed that the soul was physically located and that the "seal of the soul" could be identified.
11. For Leibniz, plastic natures were not in themselves organizing principles of the whole. For Leibniz, the monads are without windows, self-sufficient.
12. In these two articles, École showed that many "books of history of philosophy . . . have perpetuated both tenacious and regrettable misconceptions about von Wolff, for instance the idea that he only systematized and popularized Leibniz' thought."
13. Many references to Descartes show that von Wolff believed that he was following the Cartesian framework both in its theoretical content and as a program of mathematics applied to philosophy. However, it is worth asking whether Descartes would have endorsed such a deductive mode of demonstration down to the finest details.

14. It should be noted that Caspar Wolff and Christian von Wolff were not from the same family, but it is attested that Caspar Wolff studied philosophy with Christian von Wolff as his professor at the University of Halle.

Chapter Two

1. The historian Joseph Needham suggested, after Woodger's *Biological Principles* in 1929 (2014), that a turning point in the opposition between vitalism and mechanism was when "vitalist theories" postulated some entity in the living organism in addition to the chemical elements plus organizing relationships. This was opposed by what he calls "old-fashioned vitalism," which drew attention to the complexity of phenomena in contrast to the tendency of mechanists to propose oversimplified hypotheses. The new viewpoint ensures that we do not forget the extreme complexity of the living system while at the same time forbidding us to take refuge in pseudoexplanations. I argue that it is for this reason that Wolff endorses the new position of vitalism.

2. Note that this distinction between "varieties" is precisely what is called "phenotypic plasticity" in the contemporary biological literature. I discuss the emergence of this term in chapter 3.

3. Wolff is essentialist in the sense defined by Ernst Mayr, that is, he conceives the species as an essence; see Mayr (1968) and "The Biological Meaning of Species," in Mayer (1976, 515–25). As Roe (1979) has pointed out, Wolff's typological thinking explains the distinction he makes between internal cause and external form (a distinction that can only remind us of the dichotomy that will be established much later between genotype and phenotype; see chap. 3).

4. Primary ossification starts from a connective tissue (membrane) or from a cartilage blank (endochondral). Secondary ossification starts from bone tissue that has already formed (primary bone). Tertiary ossification is the permanent remodeling of bone in adults.

5. The solution to the problem of the origin of the organization is limited, since Kant makes the organization at most a regulatory concept, a presupposition.

6. This question differs from the mereological view developed by Leibniz or Christian von Wolff.

7. Except when Leibniz or von Wolff referred to the principles of optimality (the maxims of God's choice).

8. In physics, the law of elasticity states that deformation is proportional to force.

9. The term *Naturphilosophie* is German for "philosophy of nature" and refers to a school of thought that is an incarnation of German Romanticism in the scientific mode, whose aims are to search in biology for general and unifying laws governing the structure, ontogenesis, and order (diversity) of animals and plants. For a detailed analysis of this school of thought, see Richards ([2002] 2010).

10. This is the "law of fundamental biogenetics," better known as the "theory of recapitulation" and summarized by this formula. Thanks to this law, it is possible to trace the entire evolutionary history of a group of animals simply by observing the course of its embryogenesis and comparing it with that of other species. This concept is presented in E. Haeckel's book of 1866.

11. The term *plastic nature* was used by Adam Sedgwick in his *Studies from the Morphological Laboratory in the University of Cambridge* (1889).

12. I describe this notion of morphological plasticity in chapter 5.

13. In the early stages of development, the embryo of some complex organisms passes through a stage in which it has three distinct layers: the ectoderm, the mesoderm, and the endoderm.

14. *Epiboly* (from the Greek for "addition") is defined as the movement of an epithelial sheet (usually ectodermal cells) moving as a block to cover deeper structures.

15. This new experimental embryology program also benefited from newly established marine research stations, such as the famous Naples Zoological Station. Many of these experiments can only be carried out in well-equipped laboratories and in close proximity to the very diverse biological material provided by the sea.

16. Ascidians are marine animals that belong to an evolutionary group at the fringe between invertebrates and vertebrates. These animals are partly made up of tunicin, a molecule related to plant cellulose.

17. The word *evolution* (from the Latin *evolutio*, meaning "to unfold like a scroll") was originally used to describe embryological development. Its first use in connection with the development of living things dates from 1745, when Charles Bonnet used it in connection with his concept of "preformation." In the eighteenth century, the theories of Leibniz, Herder, and Schelling argued that evolution was a fundamentally spiritual process. In 1751 Pierre-Louis de Maupertuis switched to materialism. He argued that the changes that occurred during reproduction accumulated to produce new races or even new species.

18. *Neopreformationism* is a term used by Stephen J. Gould in chapter 6 of his 1977 book *Ontogeny and Phylogeny.* He uses this term to describe Wilhelm Roux and August Weismann's view on the developmental process. In the next paragraph, I resume S. J. Gould's argument.

19. *Heteroplastic* means "structurally dissimilar"; different tissues in the body—for example, nerve tissue, muscle tissue, bone tissue—are all heteroplastic in relation to each other. "As said earlier the induction effect is also possible with heteroplastic transplantation, i.e., between embryos of different species" (Spemann 1935).

20. The blastopore is the area of the embryo above the invagination area during gastrulation that, once transplanted to a ventral region, can generate a new neural plaque.

21. Like many other embryologists of his time, Spemann had little interest in the new science of genetics.

22. The blastopore is the opening in the embryo at the "blastula stage" of development that allows the archenteron (future gut) to communicate with the outside world.

23. The fibro-plastic cells are now called "fibroblasts."

24. A stem cell is an undifferentiated cell characterized by the ability to generate specialized cells though cell differentiation and the ability to maintain itself in the organism in culture by proliferation. Stem cells are present in all multicellular organisms. They play a very important role in the development of organizations and in maintaining their integrity throughout life. Leroy Stevens and Barry Pierce carried out research on mouse teratocarcinomas in the 1950s, which enabled them to isolate embryonic stem cells.

25. What I mean here by a scientifically delimited concept is one that has (1) a *birth certificate* that can be identified in the history of a field, (2) a *crucial experiment* that is historically recognizable, and (3) a *univocal definition* that persists over time.

Chapter Three

1. There are more precise distinctions between these different schools, but they don't affect our argument here.

2. Trevon Fuller, in a 2003 article entitled "The Integrative Biology of Phenotypic Plasticity," regretted that Nilsson-Ehle was "one of the figures omitted from Pigliucci's history of plasticity" (Fuller 2003, 383). Fortunately, Anthony Bradshaw, who is usually credited as the first biologist

to explicitly refer to the concept of "phenotypic plasticity," cited Nilsson-Ehle in the epigraph to his 1965 review of the literature on plasticity (I analyze Bradshaw's work and influence on plasticity in the last section of this chapter).

3. The *nature-nurture* debate is usually understood as a debate between those who support the idea that complex human traits (such as intelligence and many other traits such as personality) are largely dependent on genes and those who think that these traits are largely dependent on culture, parental care, or more generally "the environment." For more on this, see Tabery (2014).

4. The term *phenogenesis* is used to describe the development or ontogeny of a phenotype, focusing on the processes rather than the causal relationship between genotype and phenotype.

5. This type of work emerged in a specific context in which there were numerous ideological rivalries between the West and the East. On the question of learning, Watson, who was an iconoclast but in favor of behaviorism, was in the United States, while Pavlov, who was to gain international recognition, was in the East. Pavlov was seen as a model for the whole of psychology in terms of the importance of the environment in modulating behavior.

6. This phenomenon of genetic canalization is now commonly known as pleiotropy.

7. Many commentators have described Goldschmidt as one of the most controversial biologists of the twentieth century (see Dietrich 2003; Stern 1980; Sarich 1980).

8. Lysenko was a Soviet geneticist who rejected Mendelism. He was initially supported by the Soviet regime, which then sentenced him to death. He discovered the process of vernalization of wheat by exposing it to colder and colder temperatures to make it grow in winter. For more on Lysenko, see Lecourt ([1976], 1995).

Chapter Four

1. The question of the definition of the "reaction norm" and then the derived notion of the "norm of reaction," raised at the very beginning of the twentieth century by Richard Woltereck (a fervent early Darwinian), continues to be the subject of polemics to the present day (see chap. 4). Indeed, with the success of genetics in the twentieth century and its expansion through population genetics from the 1920s to the 1940s, this "tendency to variability," the causes of which Darwin sought to understand, were explained mainly in terms of genetic diversity. However, the question of the proximate causes of variation had to wait until the end of the twentieth- and the beginning of the twenty-first century, when some biologists, disappointed with the way in which the theoretical model of evolutionary biology (proposed after Darwin, in particular with the modern evolutionary synthesis) approached this question, reformulated the terms, focusing on developmental biology (see Amundson 2005). Darwin's intuitions therefore only open up a field of interest in the causes of variability in biology that continues to occupy theoretical biologists today.

2. Genecology is a branch of ecology that began in the early twentieth century and is concerned with species and varieties, their genetic subdivisions, but also their positions in nature and their ecological factors.

3. I am grateful to Laurent Loison for pointing out this episode in the history of biology, which has helped me a great deal in understanding the historical genealogy of a lingering confusion about the use of plasticity in evolutionary biology and its connection to the idea of heredity.

4. Mimicry is an adaptive strategy of imitation. Different types of mimicry strategies can be observed, such as cryptic mimicry, where a species camouflages itself in relation to the visual capabilities of the predator (or prey).

5. Rate genes have been described as genes that control the speed at which a developmental process occurs and thus indirectly control the relative efficiency of that process in competition with several other processes occurring at the same time.

6. The *collective of thought* is the term used by the French historian Stéphane Schmitt to characterize a school of thought based on the same training, ideas, and interests.

7. In short, probabilistic concepts are about the probability of occurrence, while statistical concepts are about distributions. On the difference between probabilities and statistical analysis, see Desrosières ([1993] 1998).

8. In 1947 Bernhard Rensch published a book in German that was later translated into English as *Evolution above the Species Level*. The book discusses how the evolutionary mechanisms behind speciation could also help explain differences between higher taxa.

9. George G. Simpson is an American paleontologist who has made a major contribution to the synthetic theory of evolution. In particular, he is the author of three seminal works in the discipline: *Tempo and Mode in Evolution* (1944), *The Meaning of Evolution* (1949), and *The Major Features of Evolution* (1953).

10. George L. Stebbins was an American botanist and geneticist. His major contribution to theoretical synthesis is his book *Variation and Evolution in Plants*, published in 1950.

11. The mutationist theory states that evolution has a creative force: evolution in this theory is the result of discrete mutations.

12. The Weismannian theory of germ plasma continuity states that only germ cells and their genetic components are passed on from generation to generation, whereas the somatic line is a dead end.

13. Heterozygosis is the genotypic situation where two homologous loci of the same chromosome pair each carry a different allele. Homozygosity is the presence of the same allele on both chromosomes of the same chromosome pair.

Chapter Five

1. Bradshaw thought that if "canalization" (as depicted by Waddington) was thought to have a genetic basis, then its opposite, "plasticity," was also likely to have one.

2. For Nilsson-Ehle's explicit reference to plasticity, see Nilsson-Ehle (1914).

3. This is no longer the current standard usage, which limits physiological plasticity mainly to physiological traits (such as photosynthetic rate). However, there is a current tendency to say that phenotypic plasticity should be seen as developmental plasticity (e.g., West-Eberhard 2003; Gilbert and Epel 2009; for more on this topic see Nicoglou 2013).

4. See previous note.

5. The adjective *meristematic* must be associated with the plant's meristem, a tissue that undergoes multiple cell divisions compared to the rest of the plant.

6. It should be noted that unlike Bradshaw, Nilsson-Ehle used the term *plasticity* without any reference to the adjective *phenotypic*.

7. A recent book attempts to summarize the state of the debate: *Phenotypic Plasticity and Evolution: Causes, Consequences, Controversies* (Pfennig, 2021).

8. "Vagility" in biology refers to the ability of an organism to move within an environment; such movement is called "vagile." It is therefore the ability of an organism to move freely and to migrate. Different types of vagility—such as crawling, walking, or jumping—exist for animals that move on the ground.

9. "*Plasticity* is . . . shown by a genotype when its expression is able to be altered by environmental influences"(Bradshaw 1965, 116).

10. The use of the plural "genes of plasticity" reflects the fact that by this time it had become clear that the linear causal model between a single gene and a phenotype was rarely true. Models developed in the 1990s are "polynomial" models that take into account the multiplicity of genes involved in the expression of a trait.

11. Sheiner and Lyman's view is less clear than that of De Jong (1995) and Van Tienderen (1991) because it does not identify genes of plasticity in the strict sense but rather the basis for its genetic expression.

12. The concept of overdominance dates back to the original work of Lerner (1954) and Waddington (1961). It describes the situation where the phenotype of the heterozygote is outside the phenotypic range of both homozygous parents.

13. Pleiotropy occurs when a gene influences two or more phenotypic traits.

14. Epistasis occurs when the effect of a gene mutation depends on the presence or absence of mutations in one or more other genes.

15. For an explicit test of the adaptive hypothesis of plasticity by measuring the relative fitness of alternative phenotypes in a range of environments and on the particular example of the trait "phytochrome-dependent stem elongation" in response to leaf shading using transgenic and mutant plants in which this plastic response is abolished, see Schmitt, McCormac, and Smith (1995), and Ballaré (1999).

16. In molecular genetics, an epistatic relationship generally refers to the suppression of a phenotype by another gene in the same metabolic pathway. In quantitative genetics, epistasis refers to the proportion of genetic variance that cannot be explained by the additive effects of the alleles present or by dominance effects. On the pleiotropic effects of genes and their evaluation at different phenotypic levels, see Van Tienderen, Hammad, and Zwaal (1996).

17. For a discussion of whether a gene can be considered to be the originator of a trait (i.e., genes "for" traits), see Kaplan and Pigliucci (2001).

18. John Maynard Smith (1990) was among the first to raise the issue of epigenetic inheritance.

Chapter Six

1. Contemporary philosophers of biology have recently been interested in how the explanans of the life sciences can meet the standards of general philosophy of science. Although this issue is not the focus of my argument, I will show how it helps to highlight the role of plasticity in developmental biology.

2. Schaffner argued that "most purported reductions, in biology are at best partial reductions, in which corrected or slightly modified fragments or parts of the reduced science are reduced (explained) by parts of the reducing science, and that in partial reductions a causal/mechanical approach (CM) is better at describing the results than is a formal reduction model (e.g., the GRR)" (Schaffner 2006, 384).

3. One could argue, with good reason, that another way would be to appeal not to explanatory pluralism but simply to the fact that it is possible to make distinctions. Thus, instead of distinguishing types of variation to which one would associate different types of explanation, one could simply say that there are different phenomena that share a semantic "family resemblance" (in which case there is no need to try to "unify" these phenomena).

4. *Competence* is the ability of a cell to respond to inductive signals. Not all cells are competent.

5. Other cells may receive the signal, but only competent cells can respond to it.

6. A paracrine factor has the ability to spread its signal over a relatively short distance. It has a local effect, as opposed to endocrine factors, which are hormones that can travel longer distances thanks to the circulatory system.

7. The embryo inherits a fairly compact genetic "tool kit" and uses many of the same proteins to make the heart, kidneys, teeth, eyes, and other organs. Moreover, the same proteins are used throughout the animal kingdom; for example, the factors active in the formation of the *Drosophila* eye or heart are very similar to those used in the formation of mammalian organs. Many paracrine factors can be grouped into one of four major families based on their structure: the Fibroblast Growth Factor (FGF) family, the Hedgehog family, the Wnt family, the Bone Morphogenetic Proteins (BMPs) family, and so forth.

8. Gabor Forgacs is a biologist and physician who has conducted research on cellular biophysical mechanisms during early development using both experimental approaches and modeling.

9. Blastomeres are the primordial cells of the dividing embryo. They follow the zygote.

10. A responding cell is a cell that has been induced by an inducer; a competent cell is a cell that is able to respond to that induction.

11. Gastrulation is the third stage in the development of higher metazoans and the first of morphogenesis. It consists of the establishment of the basic tissues of the embryo (or layers) by cell migration, with the differentiation of a third cell layer, the mesoblast (or mesoderm), between the two preexisting embryonic layers, the endoderm (or endoblast) and the epiblast (or exoderm or ectoblast).

12. Neurulation is a stage in the embryonic development of triploblast metazoans during which the central nervous system is formed. It follows gastrulation. Neurulation occurs in two stages: primary and secondary.

13. Triploblastic organisms are formed from three primordial layers.

Chapter Seven

1. See, for instance, Conn and Means (2000).

2. Regeneration is the ability of a living entity (cell, organ, organism, etc.) to rebuild itself after the destruction of one of its parts.

3. Cellular dedifferentiation is a characteristic of plants; some cells can return to the meristematic state and begin to divide, creating a new meristem whose activity can later give rise to a new organ. An example familiar to farmers and gardeners is that of cuttings, which allows individuals to multiply without reproducing.

4. Structural linguistics proposes to understand any language as a system in which each of the elements can be defined only by the relations of equivalence or opposition that it maintains with the others, this set of relations forming the "structure."

5. A system is understood as a set of exchanges and interactions of energy and information (the meaning is close to what was previously discussed under the term *structure*).

Chapter Eight

1. The terms *variational explanation* and *evolutionary explanation* will be used interchangeably. I note, however, that the term *evolutionary explanation* in the literature does not always refer to the variational explanation but to any explanation related to evolution.

2. This second condition allows Sober to describe children as static entities (similar to traits selected by natural selection), meaning that even when selection is not acting, the individual will always have the same trait over time.

3. For a more recent argument by Walsh for an alternative to the modern synthesis, an organism-centered view of evolution that sees organisms as adaptive agents, see Walsh, 2015.

4. It should be noted that the metaphysical position of the "selectionists" and its implications are not developed by Forber.

5. It is important to note that this quote from Dobzhansky is still the source of many discussions, mostly outside of science. The polemic arises, for example, when it comes to comparing nonscientific explanations (such as those proposed by creationists) with explanations based on natural selection.

6. To illustrate this idea of bias in the test, I refer to the famous example developed by Piaget of a six-to-seven-year-old child who is shown a bouquet of seven flowers, including five roses and two carnations and is asked whether there are more roses than flowers, or the opposite. The child who answers is mistaken because at this age she does not have the concept of "class inclusion" and therefore thinks that there are more roses than flowers, or she is confused. However, in this video (http://www.fondationjeanpiaget.ch/fjp/site/biographie/index_gen_media.php?MEDIAID =95) we can see how the little girl manages to answer the question correctly at the end of the test because the questioner, who is charitable, modifies his questions by moving them away from the strict inclusion problem, thus introducing a form of bias into the initial test.

7. Waddington used to distinguish between genetic canalization as the absence of disturbances related to genetic factors and environmental canalization as the absence of disturbances related to environmental factors. Now the term *canalization* mainly reflects what Waddington calls genetic canalization.

8. Epistatic relationships are genetic interactions between two or more genes; epistasis occurs when one or more genes mask or prevent the expression of factors at other genetic locations (loci).

9. Von Baer's laws are four: (1) during the development of the egg, the general features appear before the special features; (2) from the most general features, the less general and finally the special features would develop; (3) therefore, during its development, the organization of an animal becomes more and more different from that of other animals; (4) the embryonic stages of an animal do not resemble the adult stages of lower animals on the scale of living things but their embryonic stages. It is these last two laws that Rice questions on the basis of his observations on the evolution of canalization.

10. Massimo Pigliucci makes this explicit (2001, 68–96), but other authors interested in canalization have also made the connection between the two concepts.

11. For example, the opposition between biologists who see canalization as a direct product of natural selection and those who see canalization as a by-product of the action of selection on traits is not unlike the debate between Sara Via and Samuel Scheiner over the theoretical definition of plasticity. Sara Via, I would recall, sees plasticity as a by-product of the action of natural selection on plastic traits, while Samuel Scheiner has succeeded in measuring a certain independence between the evolution of a modulated trait and plasticity, which allows him to consider plasticity like any other trait subject to selection. The opposition seems to be the same for canalization, except that the position that canalization is only a by-product of evolution is recent and still rather marginal. In addition, most studies have focused on quantitative genetic analysis of the evolution of canalization, assuming that canalization is a trait that can be studied like any other trait.

12. Earlier in the text, Sober returns to an argument by Lewontin in which he defends the idea that evolution does not increase the fitness of the population (a point on which Sober agrees with him). From this, Lewontin concludes that the trait, once it has evolved, is not an adaptation because an adaptation must not only be the result of natural selection but must also lead to an increase in fitness (Lewontin 1978; Gould and Lewontin 1979). Sober disagrees with Lewontin on this last point, because he believes that any selected trait can be considered as an adaptation.

13. I do not enter here into the debate over the characterization of what distinguishes a causal from a noncausal explanation but simply argue that it is possible to distinguish between these two types of explanation. On the debate over the characterization of causal versus noncausal explanations, see Sober (1983). Sober believes that causal explanations trace an actual sequence of events leading to particular outcomes, whereas noncausal explanations explain an outcome by showing that a very large number of initial states of a system will evolve in the direction of the expected outcome.

Chapter Nine

1. It has been argued that evo-devo studies are not really that different from earlier modern synthetic studies in terms of the role assigned to genes except that they focus on genetic mechanisms rather than genetic frequencies.

2. In the book *Keywords and Concepts in Evolutionary Developmental Biology*, under the definition of "modularity," Gillian Gass and Jessica Bolker (2003), following von Dassow and Munro in 1999, point out that modularity has been described as "a conceptual framework for evo-devo."

3. I refer to "the evolutionary developmental synthesis" (EDS) in the title of the chapter to avoid having to provide a clear definition of evo-devo.

4. The adaptationist program is based mainly on its formulation in the 1966 book *Adaptation and Natural Selection* by G. C. Williams ([1966] 2018), as explained in chapter 6. For a recent and lively critique of selectionism and the proposal of an alternate theory of emergent evolution that is causally sufficient for evolutionary biology, see Reid (2009).

5. For a recent publication in the field, see Pfennig (2021).

6. DST proponents have defended themselves against these criticisms by producing a list of several evo-devo research agendas (Gray 2001).

Chapter Ten

1. These two aspects—form and meaning—are, of course, intrinsically linked. We distinguish them here only for the sake of clarity and precision.

2. This is reflected in the titles of More's and Cudworth's books in which they develop their interpretation of "plastic nature": More's *The Immortality of the Soul* and Cudworth's *The True Intellectual System of the Universe*. See chapter 1.

3. William James's *Principles of Psychology* first proposed the idea of neuronal plasticity in 1890. Santiago Ràmon y Cajal, who won the Nobel Prize in Medicine in 1906, theorized from this idea of plasticity, which he occasionally used. Finally, it was Geoffrey Raisman, generally considered by the scientific community to be the "father" of neural plasticity, who popularized the use of the term *neural plasticity* in an article published in 1969. This article introduced the concept of neural plasticity as the first definitive demonstration of the ability of the adult brain to form new synapses after injury.

4. There are indeed several studies that have focused on the differences between the commonly defined types of plasticity. However, their number is far less than the number of studies that insist on the similarities that exist between types of plasticity. See, for example, Novoplansky (2002, 177–88).

5. It is D. Lewis (1997) who suggests this distinction in order to clarify what dispositional properties are. Cf. Choi and Fara (2012).

6. This peak was reached first with the work of Barthez and the doctors of the Montpellier school and then especially with the work of Xavier Bichat.

7. At the beginning of the twentieth century, the embryologist Driesch developed a theory known as "neovitalist," which revived the old Aristotelian concept of "entelechy."

8. Bergson develops a metaphysical system with a vital character dedicated to the *élan vital* (vital impulse) and the *évolution créatrice* (creative evolution).

9. This accusation of Piaget and Garcia against Kuhn is somewhat exaggerated and actually refers more to the way in which the notion of paradigm has been transmitted in the philosophy of science than to the way in which it has been conceptualized by Kuhn himself. Indeed, Kuhn does not forget the epistemological dimension of the paradigm, what he calls the "disciplinary matrix" of a science. But, as we shall see, Piaget and Garcia seek to emphasize the difference between this disciplinary matrix and the epistemic framework, which is not limited to it.

10. This term *normal science* is borrowed from Kuhn. We refer to it even though Piaget and Garcia prefer the distinction between "prescience" and "science."

11. However, this observation should be qualified: authors who have the opportunity to publish in internationally circulated publications are from a rather limited number of nationalities.

12. Popper and Lakatos, for example, try to ascribe a certain rationality to the development of science in order to justify scientific progress. They therefore essentially rely on the formulation of norms that can be described as methodological.

13. For an account against reductionism, see Wimsatt (2007), who argues that our philosophy should be rooted in heuristics and models that work in practice, not just in principle.

References

Allard, Robert W., and Anda D. Bradshaw. 1964. "Implications of Genotype Environmental Interactions in Applied Plant Breeding 1." *Crop Science* 4 (5): 503–8.

Altenberg, Lee. 1994. "The Evolution of Evolvability in Genetic Programming." *Advances in Genetic Programming* 3: 47–74.

Amundson, Ron. 2005. *The Changing Role of the Embryo in Evolutionary Thought: Roots of Evo-Devo*. Cambridge: Cambridge University Press.

Aristotle. 1972. *Generation of Animals*. Leiden: E. J. Brill.

Aristotle. 2002. *On the Parts of Animals*. Oxford: Oxford University Press.

Ashby, William R. 1952. *Cybernetics: Circular Causal and Feedback Mechanisms in Biological and Social Systems; Transactions of the Eighth Conference, March 15–16, 1951*. New York: Josiah Macy Jr. Foundation.

Ashby, William R. 1956. *An Introduction to Cybernetics*. New York: John Wiley & Sons.

Aubin-Horth, Nadia, and Susan C. P. Renn. 2009. "Genomic Reaction Norms: Using Integrative Biology to Understand Molecular Mechanisms of Phenotypic Plasticity." *Molecular Ecology* 18 (18): 3763–80.

Babcock, Ernest B., and Roy Elwood Clausen. 1918. *Genetics in Relation to Agriculture*. New York: McGraw-Hill.

Bachelard, Gaston. (1938) 2002. "La formation de l'esprit scientifique: Contribution à une psychanalyse de la connaissance objective." In *The Formation of the Scientific Mind: A Contribution to a Psychoanalysis of Objective Knowledge*, translated by Mary McAllester Jones. Manchester: Clinamen.

Badyaev, Alexander V. 2005. "Stress-Induced Variation in Evolution: From Behavioural Plasticity to Genetic Assimilation." *Proceedings of the Royal Society B: Biological Sciences* 272 (1566): 877–86.

Baker, Judith R. 1974. *Race*. Oxford: Oxford University Press.

Baldwin, Mark J. 1896. "A New Factor in Evolution (Continued)." *American Naturalist* 30 (355): 536–53.

Ballaré, Carlos L. 1999. "Keeping Up with the Neighbours: Phytochrome Sensing and other Signalling Mechanisms." *Trends in Plant Science* 4 (3): 97–102.

Barrington Ernest, J. W. 1973. "Gavin Rylands de Beer." *Biographical Memoirs of Fellows of the Royal Society of London* 19:65–93.

Barthez, P.-J. *Nouveaux éléments de la science de l'homme.* Paris: Goujon, 1778.

Barton, Nick H., and Michael Turelli. 1989. "Evolutionary Quantitative Genetics: How Little Do We Know?" *Annual Review of Genetics* 23 (1): 337–70.

Bateman, K. G. 1959. "The Genetic Assimilation of Four Venation Phenocopies." *Journal of Genetics* 56:443–74.

Bateson, William. 1914. "Address of the President of the British Association for the Advancement of Science." *Science* 40 (1026): 287–302.

Baudry, Michel, Richard F. Thompson, and Joel L. Davis, eds. 1993. *Synaptic Plasticity: Molecular, Cellular, and Functional Aspects.* Cambridge, MA: MIT Press.

Beatty, John. 1982. "What's in a Word? Coming to Terms in the Darwinian Revolution." *Journal of the History of Biology* 15:215–39.

Beatty, John. (1986) 1992. "Speaking of Species: Darwin's Strategy." In *Darwin's Heritage: A Centennial Retrospect*, edited by D. Kohn. Princeton, NJ: Princeton University Press. Reprinted in M. Ereshevsky, ed., *Units of Evolution: Essays on the Nature of Species*, Cambridge, MA: MIT Press.

Beatty, John. 2006. "Chance Variation: Darwin on Orchids." *Philosophy of Science* 73 (5): 629–41.

Belsky, Jay, Charles Jonassaint, Michael Pluess, B. Brummett, and R. Williams. 2009. "Vulnerability Genes or Plasticity Genes?" *Molecular Psychiatry* 14 (8): 746–54.

Bennett, Edward L., Marian C. Diamond, David Krech, M. R. Rosenzweig, F-L. F. Chang, and W. T. Greenough. 1964. "Chemical and Anatomical Plasticity of Brain: Changes in Brain through Experience, Demanded by Learning Theories, Are Found in Experiments with Rats." *Science* 146 (3644): 610–19.

Bergmann, Carl Georg Lucas Christian, and Richard Leuckart. 1852. *Anatomisch-physiologische uebersicht des Thierreichs.* Stuttgart: J. B. Müller, 1852.

Bergson, Henri. (1938) 1998. *La pensée et le mouvant.* 13th ed. Paris: PUF.

Bergson, Henri. 1969. *L'évolution créatrice.* Paris: PUF.

Bernard, Claude. 1876. *Leçons sur la chaleur animale, sur les effets de la chaleur et sur la fièvre.* Paris: Baillière.

Blacher, L. I. 1982. *The Problem of the Inheritance of Acquired Characters.* Washington, DC: Smithsonian Institution Libraries; National Science Foundation.

Bonnet, Charles. 1745. *Traité d'insectologie.* Vol. 1. Paris: Chez Durand.

Bowler, Peter J. 1989. *The Mendelian Revolution: The Emergence of Hereditarian Concepts in Modern Science and Society.* London: Bloomsbury.

Bowler, Peter J. 2005. "Variation from Darwin to the Modern Synthesis." In *Variation*, edited by B. Hallgrímsson and B. K. Hall. Amsterdam: Elsevier Academic Press. 9–27.

Bradshaw, Anthony D. 1965. "Evolutionary Significance of Phenotypic Plasticity in Plants." *Advances in Genetics* 13:115–55.

Bradshaw, Anthony D. 2006. "Unravelling Phenotypic Plasticity: Why Should We Bother?" *New Phytologist* 170 (4): 644–48.

Brakefield, Paul M., Julie Gates, Dave Keys, F. Kesbeke, P. J. Wijngaarden, A. Montelro, V. French, and S. B. Carroll. 1996. "Development, Plasticity and Evolution of Butterfly Eyespot Patterns." *Nature* 384 (6606): 236–42.

Brigandt, Ingo. 2010. "Beyond Reduction and Pluralism: Toward an Epistemology of Explanatory Integration in Biology." *Erkenntnis* 73 (3): 295–311.

Brigandt, Ingo. 2011. "Philosophy of Biology." in *The Continuum Companion to the Philosophy of Science*, edited by S. French and J. Saatsi. London: Continuum International.

Brigandt, Ingo. 2013. "Explanation in Biology: Reduction, Pluralism, and Explanatory Aims." *Science & Education* 22 (1): 69–91.

Brigandt, Ingo, and Alan Love. 2008. "Reductionism in Biology." In *The Stanford Encyclopedia of Philosophy*, edited by E. N. Zalta and Uri Nodelman. https://plato.stanford.edu/entries/reduction-biology/.

Buffon, Georges-Louis Leclerc de. 1750. *Histoire naturelle générale et particulière, avec la description du cabinet du Roy.* Paris: Pierre de Hondt.

Burian, Richard M. 2000. "On the Internal Dynamics of Mendelian Genetics." *Comptes Rendus de l'Académie des Sciences, Series III: Sciences de la Vie* 323 (12): 1127–37.

Burns, John V. 1966. *Dynamism in the Cosmology of Christian Wolff: A Study in Pre-critical Rationalism.* New York: Exposition Press.

Callahan, Hilary S., Massimo Pigliucci, and Carl D. Schlichting. 1997. "Developmental Phenotypic Plasticity: Where Ecology and Evolution Meet Molecular Biology." *Bioessays* 19 (6): 519–25.

Calvin, William H. 1987. "The Brain as a Darwin Machine." *Nature* 330 (6143): 33–34.

Canguilhem, Georges. (1968) 1973. "Régulation." In *Encyclopaedia universalis.* Paris: Encyclopaedia universalis. https://www.universalis.fr/encyclopedie/regulation-epistemologie/.

Canguilhem, Georges 1973. "Vie." In *Encyclopaedia universalis.* Paris: Encyclopaedia universalis. https://www.universalis.fr/encyclopedie/vie/.

Canguilhem, Georges. 1981. *Idéologie et rationalité dans l'histoire des sciences de la vie: Nouvelles études d'histoire et de philosophie des sciences.* Paris: Vrin.

Carnap, Rudolph 1936. "Testability and Meaning." *Philosophy of Science* 3 (4): 419–71.

Carroll, Sean B. 2006. *Endless Forms Most Beautiful: The New Science of Evo Devo and the Making of the Animal Kingdom*, New York: W. W. Norton.

Carroll, Sean B., Jennifer K. Grenier, and Scott D. Weatherbee. (2001) 2013. *From DNA to Diversity: Molecular Genetics and the Evolution of Animal Design.* John Wiley & Sons.

Caswell, Hal. 1983. "Phenotypic Plasticity in Life-History Traits: Demographic Effects and Evolutionary Consequences." *American Zoologist* 23 (1): 35–46.

Cavalli-Sforza, Luigi L. 1974. "The Genetics of Human Populations." *Scientific American* 231 (3): 80–91.

Cheverud, James M., James J. Rutledge, and William R. Atchley. 1983. "Quantitative Genetics of Development: Genetic Correlations Among Age-Specific Trait Values and the Evolution of Ontogeny." *Evolution* 37 (5): 895–905.

Child, Charles M. 1902. "Studies on Regulation." *Archiv für Entwicklungsmechanik der Organismen* 15 (2): 187–237.

Child, Charles M. 1903. "Studies on Regulation." *Archiv für Entwicklungsmechanik der Organismen* 17 (1): 1–40.

Child, Charles M. 1903. "Studies on Regulation." *Archiv für Entwicklungsmechanik der Organismen* 20 (1): 48–75.

Child, Charles M. 1906. "Contributions toward a Theory of Regulation." *Archiv für Entwicklungsmechanik der Organismen* 20 (3): 380–426.

Choi, Sungho. 2005. "Dispositions and Mimickers." *Philosophical Studies* 122:183–88.

Choi, Sungho. 2011. "Finkish Dispositions and Contextualism." *Monist* 94 (1): 103–20.

Choi, Sungho, and Michael Fara. 2012. "Dispositions." In *The Stanford Encyclopedia of Philosophy*, edited by E. N. Zalta and Uri Nodelman. http://plato.stanford.edu/archives/spr2012/entries/dispositions/.

Coleman, William. 1980. "Morphology in the Evolutionary Synthesis." In *The Evolutionary Synthesis: Perspectives on the Unification of Biology*, edited by Ernst Mayr and W. B. Provine, 174–80. Cambridge, MA: Harvard University Press.

Comte, Auguste. 1876. *Course of Positive Philosophy*. London. George Bell and Sons.

Conn, P. Michael, and Anthony R. Means, eds. 2000. *Principles of Molecular Regulation*. Totowa, NJ: Humana Press.

Cooper, John M. 1988. "Metaphysics in Aristotle's Embryology." *Cambridge Classical Journal* 34:14–41.

Corr, Charles A. 1972. "Christian Wolff's Treatment of Scientific Discovery." *Journal of the History of Philosophy* 10 (3): 323–34.

Corr, Charles A. 1975. "Christian Wolff and Leibniz." *Journal of the History of Ideas* 36 (2): 241–62.

Crespi, Erica J., and Robert J. Denver. 2005. "Roles of Stress Hormones in Food Intake Regulation in Anuran Amphibians Throughout the Life Cycle." *Comparative Biochemistry and Physiology Part A: Molecular & Integrative Physiology* 141 (4): 381–90.

Cudworth, Ralph. (1678) 1829. *The Works of Ralph Cudworth: Containing the True Intellectual System of the Universe, Sermons, &c.* Vol. 1. Oxford: D. A. Talboys.

Darwin, Charles. 1859. *On the Origin of Species by Means of Natural Selection, or, the Preservation of Favoured Races in the Struggle for Life*. London: John Murray.

Darwin, Charles. 1861. *On the Origin of Species by Means of Natural Selection, or, the Preservation of Favoured Races in the Struggle for Life*. 3rd ed. London: John Murray.

Darwin, Charles. 1876. *On the Origin of Species by Means of Natural Selection, or, the Preservation of Favoured Races in the Struggle for Life*. 6th ed. London: John Murray.

Darwin, Charles. (1868) 1872. *The Variation of Animals and Plants under Domestication*. 2nd ed. London: John Murray.

Dawkins, Richard. (1982) 2016. *The Extended Phenotype: The Long Reach of the Gene*. Oxford: Oxford University Press.

De Beer, Gavin, R. 1930. *Embryology and Evolution*. Oxford: Clarendon Press.

De Beer, Gavin, R. 1940. *Embryos and Ancestors*. Oxford: Clarendon Press.

De Beer, Gavin, R. and Stephen Goodrich. 1938. *Evolution Essays on Aspects of Evolutionary Biology, Presented to Professor E.S. Goodrich*. Oxford: Clarendon Press.

De Jong, Gerdien. 1995. "Phenotypic Plasticity as a Product of Selection in a Variable Environment." *American Naturalist* 145 (4): 493–512.

De Jong, Gerdien. 2005. "Evolution of Phenotypic Plasticity: Patterns of Plasticity and the Emergence of Ecotypes." *New Phytologist* 166 (1): 101–18.

Descartes, René. (1641) 2016. "Meditations on First Philosophy." In *Seven Masterpieces of Philosophy*, edited by Steven M. Cahn. Abingdon: Routledge. 63–108.

Descartes, René. (1649) 2017. *The Passions of the Soul*. https://www.earlymoderntexts.com/assets/pdfs/descartes1649.pdf.

Descartes, René, and D. Antoine-Mahut. (1677) 2016. *Treatise on Man and Its Reception*. Cham, Switzerland: Springer.

Desrosières, Alain. (1993) 1998. *The Politics of Large Numbers: A History of Statistical Reasoning*. Cambridge, MA: Harvard University Press, 1998.

De Visser, Arjan J., G. M. Joachim Hermisson, Günter P. Wagner, Lauren Ancel Meyers, Homayoun Bagheri-Chaichian, Jeffrey L Blanchard, Lin Chao, James M. Cheverud, Santiago F. Elena, and Walter Fontana. 2003. "Perspective: Evolution and Detection of Genetic Robustness." *Evolution* 57 (9): 1959–72.

DeWitt, Thomas J., and Samuel M. Scheiner, eds. 2004. *Phenotypic Plasticity: Functional and Conceptual Approaches*. Oxford: Oxford University Press.

Dietrich, Michael R. 2003. "Richard Goldschmidt: Hopeful Monsters and Other 'Heresies.'" *Nature Reviews Genetics* 4 (1): 68–74.

Dingemanse, Niels J., Anahita J. N. Kazem, D. Réale, and Jonathan Wright. 2010. "Behavioural Reaction Norms: Animal Personality Meets Individual Plasticity." *Trends in Ecology & Evolution* 25, (2): 81–89.

Dobzhansky, Theodosius. 1926. "K Voprosu o Nasledovanii Priobetennykh Priznakov" [On the question of the heredity of acquired characteristics]. In *Preformizm Ili Epigenezis?* [Preformation or epigenesis], edited by E. S. Smirnov and N. D. Leonov, 27–47. Vologda: Timiriazev Institute.

Dobzhansky, Theodosius. (1937) 1982. *Genetics and the Origin of Species*. Columbia Classics in Evolution Series, no. 11. New York: Columbia University Press.

Dobzhansky, Theodosius. 1955. *Evolution, Genetics, and Man*. New York: Wiley.

Dobzhansky, Theodosius. (1973) 2013. "Nothing in Biology Makes Sense Except in the Light of Evolution." *American Biology Teacher* 75 (2): 87–91.

Dobzhansky, Theodosius, and Ernest Boesiger. 1983. *Human Culture: A Moment in Evolution*. Edited by B. Wallace. New York: Columbia University Press.

Dobzhansky, Theodosius, and Bruce Wallace. 1953. "The Genetics of Homeostasis in Drosophila." *Proceedings of the National Academy of Sciences* 39 (3): 162–71.

Donohue, Kathleen. 2005. "Niche Construction through Phenological Plasticity: Life History Dynamics and Ecological Consequences." *New Phytologist* 166 (1): 83–92.

Driesch, Hans. (1914) 2010. *The History and Theory of Vitalism*. London: Macmillan.

Duchesneau, François. 1998. *Les modèles du vivant de Descartes à Leibniz* [Models of the living beings from Descartes to Leibniz]. Paris. Vrin.

Dunn, Leslie Clarence. (1965) 1991. *A Short History of Genetics: The Development of Some of the Main Lines of Thought 1864–1939*. Ames: Iowa State University Press.

Dupont, Jean-Claude. 2007. "Pre-Kantian Revival of Epigenesis: Caspar Friedrich Wolff's, *De formation intestinorum*." In *Understanding Purpose: Kant and the Philosophy of Biology*, edited by P. Huneman. Rochester, NY: University Rochester Press.

Dupont, Jean-Claude, and Stéphane Schmitt. 2004. *Du feuillet au gène: Une histoire de l'embryologie moderne, fin XVIIIe–XXe siècle* [From layers to genes: A history of modern embryology, end of 18th–20th century]. Paris: Rue d'Ulm.

Dupré, John. 1993. *The Disorder of Things: Metaphysical Foundations of the Disunity of Science*. Cambridge, MA: Harvard University Press.

École, Jean. 1964. "*Cosmologie wolffienne et dynamique leibnizienne: Essai sur les rapports de Wolff avec Leibniz*" [Wolffian cosmology and Leibnizian dynamics: Essai on the relationships of Wolff with Leibniz]. *Les études philosophiques* 19 (1): 3–9.

École, Jean. 1979."*En quels sens peut-on dire que Wolff est rationaliste?*" [In what sense can Wolff be said to be a rationalist?]. *Studia Leibnitiana* 11 (1): 45–61.

Eshel, Ilan, and Carlo Matessi. 1998. "Canalization, Genetic Assimilation and Preadaptation: A Quantitative Genetic Model." *Genetics* 149 (4): 2119–33.

Ettensohn, Charles A., C. Kitazawa, M. S. Cheers, J. D. Leonard, and T. Sharma. 2007. "Gene Regulatory Networks and Developmental Plasticity in the Early Sea Urchin Embryo: Alternative Deployment of the Skeletogenic Gene Regulatory Network." *Development* 134:3077–87.

Falconer, Douglas S. 1952. "The Problem of Environment and Selection." *American Naturalist* 86 (830): 293–98.

Falconer, Douglas S. (1960) 1981. *Introduction to Quantitative Genetics*. 2nd ed. London: Longman.

Faucher, Luc, Pierre Poirier, and Jean Lachapelle. 2006. "La théorie des systèmes développementaux et la construction sociale des maladies mentales" [Developmental systems theory and the social construction of mental illnesses]. *Philosophiques* 33(1): 147–82.

Fischer, Jean-Louis. 2002a. "Créations et fonctions des stations maritimes françaises" [Creation and functions of French maritime stations]. *La revue pour l'histoire du CNRS* 7:26–31.

Fischer, Jean-Louis. 2002b. "Les origines de l'embryologie expérimentale et les nouvelles formulations théoriques de la néo-épigenèse et de la néo-préformation à la fin du XIXe siècle" [The origins of experimental embryology and the new theoretical formulations of neoepigenesis and neopreformation at the end of the nineteenth century]. *Bulletin d'histoire et d'épistémologie des sciences de la vie* 9:147–62.

Fisher, Mark. 2007. "Generation and Classification of Organisms in Kant's Natural Philosophy." In *Understanding Purpose: Kant and the Philosophy of Biology*, edited by P. Huneman, 101–21. Rochester, NY: University Rochester Press.

Fisher, Ronald A. 1918. "The Correlation between Relatives on the Supposition of Mendelian Inheritance." *Earth and Environmental Science Transactions of the Royal Society of Edinburgh* 52 (2): 399–433.

Fisher, Ronald A. (1930) 1999. *The Genetical Theory of Natural Selection: A Complete Variorum Edition*. Oxford: Oxford University Press, 1999.

Fitter, Alastair H. 2010. "Anthony David Bradshaw: 17 January 1926—21 August 2008." *Biographical Memoirs of Fellows of the Royal Society* 56:25–39.

Flatt, Thomas. 2005. "The Evolutionary Genetics of Canalization." *Quarterly Review of Biology* 80 (3): 287–316.

Foehring, Robert C., and Nancy M. Lorenzon. 1999. "Neuromodulation, Development and Synaptic Plasticity." *Canadian Journal of Experimental Psychology/Revue canadienne de psychologie expérimentale* 53 (1): 45–61.

Forber, Patrick. 2005. "On the Explanatory Roles of Natural Selection." *Biology and Philosophy* 20 (2/3): 329–42.

Forgacs, Gabor, and Stuart A. Newman. 2005. *Biological Physics of the Developing Embryo*. Cambridge: Cambridge University Press.

Friml, Jiri, and Michael Sauer. 2008. "In Their Neighbour's Shadow." *Nature* 453 (7193): 298–99.

Fuller, Trevon. 2003. "The Integrative Biology of Phenotypic Plasticity." *Biology and Philosophy* 18 (2): 381–89.

Furusawa, Chikara, and Kunihiko Kaneko. 2006. "Morphogenesis, Plasticity and Irreversibility." *International Journal of Developmental Biology* 50 (2/3): 223–32.

Futuyma, Douglas. J. 2005. *Evolution*. Sunderland, MA: Sinauer Associates.

Galliot, Brigitte, M. Miljkovic-Licina, R. de Rosa, and S. Chera. 2006. "Hydra, a Niche for Cell and Developmental Plasticity." *Seminars in Cell & Developmental Biology* 17 (4): 492–502.

Gass, Gillian L., and Bolker, Jessica A. 2003. "Modularity." *Keywords and Concepts in Evolutionary Developmental Biology*, edited by Brian K. Hall and Wendy M. Olson, 260–67. Cambridge, MA: Harvard University Press.

Gassendi, Pierre. 1658. *Opera omnia*. Lyon: L. Anisson & Joan B. Devenet.

Gause, Georgii F. 1936. "The Principles of Biocoenology." *Quarterly Review of Biology* 11 (3): 320–36.

Gause, Georgii F. 1947. "Problems of Evolution." *Transaction of the Connecticut Academy of Sciences* 37:17–68.

Gayon, Jean. (1992) 2019. *Darwin et l'aprés-Darwin: Une histoire de l'hypothèse de selection na-turelle* [Darwin and after Darwin: A history of the hypothesis of natural selection]. Ed. rev. and corr. by Françoise Parot. Paris: Éditions Matériologiques.

Gayon, Jean. 1998. *Darwin's Struggle for Survival: Heredity and the Hypothesis of Selection.* Cambridge Studies in Philosophy and Biology. Cambridge: Cambridge University Press.

Gilbert, Scott F. 1988. "Cellular Politics: Ernest Everett Just, Richard B. Goldschmidt, and the Attempt to Reconcile Embryology and Genetics." In *The American Development of Biology,* edited by R. Rainger, K. Benson, and J. Maienschein, 311–46. Philadelphia. University of Pennsylvania Press.

Gilbert, Scott F. 2000a. *Developmental Biology.* 6th ed. Sunderland, MA: Sinauer Associates.

Gilbert, Scott F. 2000b. "Diachronic Biology Meets Evo-Devo: CH Waddington's Approach to Evolutionary Developmental Biology." *American Zoologist* 40 (5): 729–37.

Gilbert, Scott F. 2001. "Ecological Developmental Biology: Developmental Biology Meets the Real World." *Developmental biology* 233 (1): 1–12.

Gilbert, Scott F. 2006. *Developmental Biology.* 8th ed. Sunderlund, MA: Sinauer Associates.

Gilbert, Scott F., and David Epel. 2009. *Ecological Developmental Biology: Integrating Epigenetics, Medicine, and Evolution.* Sunderland, MA. Sinauer Associates.

Gilbert, Scott F., and Michael J. F. Barresi. 2016. *Developmental Biology.* 11th ed. Sunderland, MA: Sinauer Associates.

Gillespie, John H., and Michael Turelli. 1989. "Genotype-Environment Interactions and the Maintenance of Polygenic Variation." *Genetics* 121 (1): 129–38.

Goldschmidt, Richard. 1940. *The Material Basis of Evolution.* Yale, CT. Yale University Press.

Gomulkiewicz, Richard, and Mark Kirkpatrick. 1992. "Quantitative Genetics and the Evolution of Reaction Norms." *Evolution* 46, no. 2): 390–411.

Gotthard, Karl, and Sören Nylin. 1995. "Adaptive Plasticity and Plasticity as an Adaptation: A Selective Review of Plasticity in Animal Morphology and Life History." *Oikos* 74 (1): 3–17.

Gould, Stephen J. 1977. *Ontogeny and Phylogeny.* Cambridge, MA: Harvard University Press.

Gould, Stephen Jay, and Richard C. Lewontin. 1979. "The Spandrels of San Marco and the Panglossian Paradigm: A Critique of the Adaptationist Programme." *Proceedings of the Royal Society of London B* 205 (1161): 581–98.

Gray, Russell D. 2001. "Selfish Genes or Developmental Systems?" *Thinking about Evolution: Historical, Philosophical, and Political Perspectives,* edited by Rama S. Singh, 184–207. Cambridge University Press.

Grene, Marjorie. 1974. *The Understanding of Nature: Essays in the Philosophy of Biology.* Vol. 23. The Netherlands: Springer.

Griffiths, Paul E., and Russell D. Gray. 1994. "Developmental Systems and Evolutionary Explanation." *Journal of Philosophy* 91 (6): 277–304.

Griffiths, Paul E., and Robin D. Knight. 1998. "What Is the Developmentalist Challenge?" *Philosophy of Science* 65 (2): 253–58.

Haeckel, Ernst. 1866. *Generelle Morphologie der Organismen: Ellgemeine Grundzüge des organischen Formen-Wissenschaft, mechanisch begründet durch die von Charles Darwin reformirte Descendenz-Theorie.* Berlin: Reimer.

Hall, Brian K. (1992) 1999. *Evolutionary Developmental Biology.* 2nd ed. The Netherlands: Springer.

Hall, Brian K. 2000. "Balfour, Garstang and De Beer: The First Century of Evolutionary Embryology." *American zoologist* 40 (5): 718–28.

Hall, Brian K. 2003. "Unlocking the Black Box between Genotype and Phenotype: Cell Condensations as Morphogenetic (Modular) Units." *Biology and Philosophy* 18 (2): 219–47.

Hall, Brian K. 2005. "Consideration of the Neural Crest and Its Skeletal Derivatives in the Context of Novelty/Innovation." *Journal of Experimental Zoology Part B: Molecular and Developmental Evolution* 304 (6): 548–57.

Hall, Brian K., and Wendy M. Olson, eds. 2006. *Keywords and Concepts in Evolutionary Developmental Biology*. Cambridge, MA: Harvard University Press.

Hallgrímsson, Benedikt, and Brian K. Hall. 2005. "Variation and Variability: Central Concepts in Biology." In *Variation*, edited by Benedikt Hallgrímsson and Brian K. Hall, 1–7. Amsterdam: Elsevier Academic Press.

Hamburger, Viktor. (1980) 1998. "Embryology and the Modern Synthesis in Evolutionary Theory." In *The Evolutionary Synthesis*, edited by E. Mayr and W. Provine, 97–112. Cambridge, MA: Harvard University Press.

Harper, John L. 1967. "A Darwinian Approach to Plant Ecology." *Journal of Applied Ecology* 4 (2): 267–90.

Harvey, William. 1651. *Exercitationes de Generatione Animalium*. London. Typus DU-Gardinas; Impensis Octaviani Pulleyn.

Harvey, William. 1628. *Exercitatio Anatomical de Motu Cordis et Sanguinis in Animalibus*. Translated by H. E. Sigerist. Frankfurt: Sumptibus Guilielmi Fitzeri.

Harvey, William. 1847. *The Works of William Harvey, MD*. Translated by R. Willis. London: Sydenham Society.

Hempel, Carl G. 1966. *Philosophy of Natural Science*. Englewood Cliffs, NJ: Prentice-Hall.

Hettche, Matt, and Corey Dyck. 2019. "Christian Wolff." *The Stanford Encyclopedia of Philosophy*, edited by E. N. Zalta and Uri Nodelman. https://plato.stanford.edu/archives/win2019/entries/wolff-christian/.

His, William. 1887. *Über die Methoden der plastischen Rekonstruction und über deren Bedeutung für Anatomie U. Entwicklungsgeschichte*. Jena: Frommansche Buchdruckerei; Hermann Pohle.

Hogben, Lancelot. 1933. *Nature and Nurture*. New York: W. W. Norton.

Huneman, Philippe. 2007. "Reflexive Judgement and Wolffian Embryology: Kant's Shift between the First and the Third Critique." In *Understanding Purpose? Kant and the Philosophy of Biology*, edited by Philippe Huneman, 75–100. Rochester, NY: University of Rochester Press.

Huneman, Philippe. 2014. "Formal Darwinism as a Tool for Understanding the Status of Organisms in Evolutionary Biology." *Biology & Philosophy* 29:271–79.

Huneman, Philippe, and Anne-Lise Rey. 2007. "La controverse Leibniz-Stahl dite Negotium otiosum." *Bulletin d'Histoire et d'épistémologie des sciences de la vie* 14 (2): 213–38.

Hunter, William B. 1950. "The Seventeenth Century Doctrine of Plastic Nature." *Harvard Theological Review* 43 (3):197–213.

Huxley, Julian, and Gavin de Beer. 1934. *The Elements of Experimental Embryology*. Cambridge: Cambridge University Press.

Huxley, J. S., and E. B. Ford. 1925. "Mendelian Genes and Rates of Development." *Nature* 116 (2928): 861–63.

Hyde, Janet S. 1973. "Genetic Homeostasis and Behavior: Analysis, Data, and Theory." *Behavior Genetics* 3, no. 3): 233–45.

Inhelder, Bärbel, and Jean Piaget. (1959) 1991. *La genèse des structures logiques élémentaires: Classifications et sériations*. 5th ed. Neuchâtel: Delachaux & Niestlé.

Jablonka, Eva, and Marion J. Lamb, and A. Zeligowski. (2006) 2014. *Evolution in Four Dimensions: Genetic, Epigenetic, Behavioral, and Symbolic Variation in the History of Life.* Rev. ed. Cambridge, MA: MIT Press.

Jacob, François. (1981) 2022. *The Logic of Life: A History of Heredity.* Vol. 62. Princeton, NJ: Princeton University Press.

Jacob, François. 1998. *Of Flies, Mice, and Men.* Cambridge, MA: Harvard University Press.

Jacob, François, and Jacques Monod. 1961. "On the Regulation of Gene Activity." *Cold Spring Harbor Symposia on Quantitative Biology* 26:193–211.

Jain, S. K., and Anda D. Bradshaw. 1966. "Evolutionary Divergence among Adjacent Plant Populations I. The Evidence and Its Theoretical Analysis." *Heredity* 21 (3): 407–41.

James, William. 1890. *Principles of Psychology.* New York: Henry Holt.

Jenkinson, John Wilfrid. 1909. *Experimental Embryology.* Oxford: Clarendon Press, 1909.

Jinks, John Leonard, and Kenneth Mather. 1955. "Stability in Development of Heterozygotes and Homozygotes." *Proceedings of the Royal Society of London B: Biological Sciences* 143 (913): 561–78.

Jinks, J. L., and H. S. Pooni. 1988. "The Genetic Basis of Environmental Sensitivity." *Proceeding of the Second International Conference on Quantitative Genetics,* 505–22. Sunderland, MA: Sinaure Associates.

Johannsen, Wilhelm. 1903. "Om arvelighed I samfund og i rene linier" [About heredity in communities and in pure lines]. *Oversigt over det Kongelige Danske Videnskabernes Selskabs Forhandlinger* 3: 247–70.

Johannsen, Wilhelm. 1905. *Arvelighedsloerens Elementer* [The elements of heredity]. Copenhagen: Gyldendal.

Johannsen, Wilhelm. (1909) 1913. *Elemente der exakten Erblichkeitslehre* [The elements of exact heredity]. Jena: G. Fischer.

Johannsen, Wilhelm. 1911. "The Genotype Conception of Heredity." *American Naturalist* 45 (531): 129–59.

Kaplan, Jonathan Michael, and Massimo Pigliucci. 2001. "Genes 'for' Phenotypes: A Modern History View." *Biology and Philosophy* 16 (2): 189–213.

Kimura, Motoo. 1955. "Stochastic Processes and Distribution of Gene Frequencies under Natural Selection." *Cold Spring Harbor Symposia on Quantitative Biology* 20 (1): 33–53.

Kirkpatrick, Mark, and Nancy Heckman. 1989. "A Quantitative Genetic Model for Growth, Shape, Reaction Norms, and Other Infinite-Dimensional Characters." *Journal of Mathematical Biology* 27: 429–50.

Kitcher, Philip. 1989. "Explanatory Unification and the Causal Structure of the World." In *Scientific Explanation,* edited by P. Kitcher and W. Salmon. Minneapolis: University of Minnesota Press.

Krafka, Joseph, Jr. 1920. "The Effect of Temperature upon Facet Number in the Bar-Eyed Mutant of Drosophila: Part I." *Journal of General Physiology* 2 (5): 433–44.

Kuhn, Thomas S. (1962) 2012. *The Structure of Scientific Revolutions.* Chicago: University of Chicago Press.

Lamarck, Jean-Baptiste (de). 1809. *Philosophie zoologique ou exposition des considérations relatives à l'histoire naturelle des animaux.* Paris. Dentu.

Lande, Russell. 1980. "Genetic Variation and Phenotypic Evolution during Allopatric Speciation." *American Naturalist* 116 (4): 463–79.

Lande, Russell, and Stevan J. Arnold. 1983. "The Measurement of Selection on Correlated Characters." *Evolution* 37 (6): 1210–26.

Laplane, Lucie. 2016. *Cancer Stem Cells: Philosophy and Therapies*. Cambridge, MA: Harvard University Press.

Larsen, Ellen, Tim Lee, and Nathalia Glickman. 1996. "Antenna to Leg Transformation: Dynamics of Developmental Competence." *Developmental genetics* 19 (4): 333–39.

Laubichler, Manfred D., and Jane Maienschein, eds. 2009. *From Embryology to Evo-Devo: A History of Developmental Evolution*. Cambridge, MA: MIT Press.

Lauder, George V. 1982. "Introduction." In *Form and Function: A Contribution to the History of Animal Morphology*, edited by E. S. Russell, xi–xlv. Chicago: Chicago University Press.

Laurence, John Zachariah. 1855. *The Diagnosis of Surgical Cancer: The Liston Prize Essay for 1854*. London: J. Churchill.

Lawrence, Cera R. 2008. "Ovism." *Embryo Project Encyclopedia*. KEEP, Arizona State University, August 8. https://hdl.handle.net/10776/1801.

Lecourt, Dominique. (1976) 1995. *Lyssenko: Histoire réelle d'une "science prolétarienne"* [Lysenko: The real story of a "proletarian science]. Paris. Quadrige/PUF.

Lecuit, Thomas, and Pierre-Francois Lenne. 2007. "Cell Surface Mechanics and the Control of Cell Shape, Tissue Patterns and Morphogenesis." *Nature Reviews Molecular Cell Biology* 8 (8: 633–44.

Leibniz, Gottfried W. (1705) 1840. *Considération sur les principes de vie et sur les natures plastiques*. Vol. 2, pt. 1. Paris: Ed. Dutens.

Leibniz, Gottfried W. (1714) 2014. *Leibniz's Monadology: A New Translation and Guide*. Translated by Lloyd Strickland. Edinburgh: Edinburg University Press.

Lenders, Winfried. 1971. "The Analytic Logic of GW Leibniz and Chr. Wolff: A Problem in Kant Research." *Synthese* 23 (1): 147–53.

Lerner, Isadore M. 1954. *Genetic homeostasis*. London. Oliver & Boyd.

Lerner, Richard M. 1984. *On the Nature of Human Plasticity*. Cambridge: Cambridge University Press.

Levins, Richard. 1963. "Theory of Fitness in a Heterogeneous Environment. II. Developmental Flexibility and Niche Selection." *American Naturalist* 97 (893): 75–90.

Lewis, David. 1997. "Finkish Dispositions." *Philosophical Quarterly* 47 (187): 143–58.

Lewontin, Richard C. 1957. "The Adaptations of Populations to Varying Environments." *Cold Spring Harbor Symposia on Quantitative Biology* 22 (1): 395–408.

Lewontin, Richard C. 1978. "Adaptation." *Scientific American* 239 (3): 212–28.

Lewontin, Richard. 1983. "Darwin's Revolution." *New York Review of Books* 30 (10): 34–37.

Lively, Curtis M. 1986. "Predator-Induced Shell Dimorphism in the Acorn Barnacle *Chthamalus anisopoma*." *Evolution* 40 (2): 232–42.

Loison, Laurent. 2010. *Qu'est-ce que le néolamarckisme: Les biologistes français et la question de l'évolution des espèces, 1870–1940*. Paris: Vuibert.

Lukin, Efim J. 1936. "On the Causes of Substitution of Modifications by Mutations in the Process of Organic Evolution From the Viewpoint of the Theory of Natural Selection" [in Ukrainian]. *Trans. Kharhov University* 6: 199–209.

Lynch, Michael, and Wilfried Gabriel. 1987. "Environmental Tolerance." *American Naturalist* 129 (2): 283–303.

Maienschein, Jane. 2003. *Whose View of Life?: Embryos, Cloning, and Stem Cells*. Cambridge, MA: Harvard University Press.

Malpighi, Marcello. 1673. *Dissertatio Epistolica de Formatione Pulli in Ovo* [The study of the chick embryo]. London.

Marriott, Cliff G., and Graham J. Holloway. 1998. "Colour Pattern Plasticity in the Hoverfly, *Episyrphus balteatus*: The Critical Immature Stage and Reaction Norm on Developmental Temperature." *Journal of Insect Physiology* 44 (2): 113–19.

Mather, Kenneth. 1953. "Genetical Control of Stability in Development." *Heredity* 7 (3): 297–336.

Maupertuis, Pierre-Louis M. de. (1745) 2012. *Vénus physiques*. Paris: Hachette Livre Bnf.

Mayr, Ernst. (1960) 1997. *Evolution and the Diversity of Life: Selected Essays*. 5th ed. Cambridge, MA: Harvard University Press.

Mayr, Ernst. 1968. "Illiger and the Biological Species Concept." *Journal of the History of Biology* 1 (2): 163–78.

Mayr, Ernst. 1982. *The Growth of Biological Thought: Diversity, Evolution, and Inheritance*. Cambridge, MA: Harvard University Press.

Meckel, Johann F. 1827. *Archiv für Anatomie und Physiologie*. Vol. 12. Leipzig: Leopold Voss.

Meckel, Johann F. 1830. *Anatomie und Physiologie*. Leipzig: Leopold Voss.

Mehnert, Ernst. 1898. *Biomechanik, erschlossen aus dem Principe der Organogenese*. Jena: Fischer.

Metcalf, Maynard M. 1906. "The Influence of the Plasticity of Organisms upon Evolution." *Science* 23 (594): 786–87.

Metcalf, Maynard M. 1906. *An Outline of the Theory of Organic Evolution: With a Description of Some of the Phenomena Which It Explains*. New York: Macmillian.

Metlen, Kerry L., Erik T. Aschehoug, and Ragan M. Callaway. 2009. "Plant Behavioural Ecology: Dynamic Plasticity in Secondary Metabolites." *Plant, Cell & Environment* 32 (6): 641–53.

Mitchell, Richard S. 1976. "Submergence Experiments on Nine Species of Semi-Aquatic Polygonum." *American Journal of Botany* 63 (8): 1158–65.

Mitchell, Sandra D. 2003. *Biological Complexity and Integrative Pluralism*. Cambridge, MA: Cambridge University Press.

Morange, Michel. 2012. *Les secrets du vivant: Contre la pensée unique en biologie*. Paris: La Découverte.

More, Henry. 1659. *The Immortality of the Soul: So Farre Forth as it is Demonstrable from the Knowledge of Nature and the Light of Reason*. London: J. Flesher.

Morgan, Conwy L. 1900. *Animal Behaviour*. London: E. Arnold.

Morgan, Thomas H. 1917. "The Theory of the Gene." *American Naturalist* 51, no. 609: 513–44.

Müller-Wille, Staffan. 2005. "Early Mendelism and the Subversion of Taxonomy: Epistemological Obstacles as Institutions." *Studies in History and Philosophy of Science Part C: Studies in History and Philosophy of Biological and Biomedical Sciences* 36 (3): 465–87.

Müller, Gerd B., and Stuart A. Newman, eds. 2003. *Origination of Organismal Form: Beyond the Gene in Developmental and Evolutionary Biology*. Vol. 2. Cambridge, MA: MIT Press.

Müller, Gerd B., and Stuart A. Newman. 2005. "The Innovation Triad: An Evodevo Agenda." *Journal of Experimental Zoology Part B: Molecular and Developmental Evolution* 304 (6): 487–503.

Murray, Stacey R. 2000. *John Tuberville Needham: An Entry from Gale's Science and Its Times*. Farmington Hills, MI: Gale.

Nanjundiah, Vidyanand. 2003. "Phenotypic Plasticity and Evolution by Genetic Assimilation." In *Origination of Organismal Form: Beyond the Gene in Developmental and Evolutionary Biology*, edited by Gerd Müller, 245–63. Cambridge, MA: MIT Press 245–63.

Naudin, Charles. 1852. "Considérations philosophiques sur l'espèce et la variété." [Philosophical considerations on species and varieties]. *Revue horticole* 4 (1): 102–9.

Neander, Karen. 1988. "Discussion: What Does Natural Selection Explain? Correction to Sober."
   Philosophy of Science 55 (3): 422–26.

Needham, John T. 1747. *New Microscopical Discoveries*. Leiden: A. Trembley.

Needham, Joseph. (1968) 2015. *Order and Life*. Cambridge: Cambridge University Press.

Needham, Joseph, and Arthur Hughes. (1959) 2015. *A History of Embryology*. Cambridge: Cambridge University Press.

Newman, Stuart A. 1992. "Generic Physical Mechanisms of Morphogenesis and Pattern Formation as Determinants in the Evolution of Multicellular Organization." *Journal of Biosciences* 17: 193–215.

Newman, Stuart A. 2010. "Dynamical Patterning Modules." In *Evolution: The Extended Synthesis*, edited by Massimo Pigliucci and Gerd B. Müller, 281–306. Cambridge, MA: MIT Press.

Newman, Stuart A. 2011. "Animal Egg as Evolutionary Innovation: A Solution to the 'Embryonic Hourglass' Puzzle." *Journal of Experimental Zoology Part B: Molecular and Developmental Evolution* 316 (7): 467–83.

Newman, Stuart A. 2023. "Inherency and Agency in the Origin and Evolution of Biological Functions." *Biological Journal of the Linnean Society* 139 (4): 487–502.

Newman, Stuart A., Ramray Bhat, and Nadejda V. Mezentseva. 2009. "Cell State Switching Factors and Dynamical Patterning Modules: Complementary Mediators of Plasticity in Development and Evolution." *Journal of Biosciences* 34 (4): 553–72.

Nicholson, Alexander J. 1954. "An Outline of the Dynamics of Animal Populations." *Australian Journal of Zoology* 2 (1): 9–65.

Nicholson, Alexander John. 1957. "The Self-Adjustment of Populations to Change." *Cold Spring Harbor Symposia on Quantitative Biology* 22:153–73.

Nicoglou, Antonine. 2013. "West Eberhard and the Notion of Plasticity." In *Proceedings of the CAPE International Workshops, 2012*. CAPE Studies in Applied Philosophy and Ethics, vol. 1, 26–38. Kyoto: CAPE.

Nicoglou, Antonine. 2017. "The Timing of Development." In *Time of Nature and the Nature of Time: Philosophical Perspectives of Time in Natural Sciences*, edited by C. Bouton and P. Huneman, 359–90. Cham, Switzerland: Springer.

Nicoglou, Antonine. 2018. "Waddington's Epigenetics or the Pictorial Meetings of Development and Genetics." *History and Philosophy of the Life Sciences* 40: 1–25.

Nieuwkoop, Pieter D., Anna Gisela Johnen, and Brigitte Albers. 1985. *The Epigenetic Nature of Early Chordate Development: Inductive Interaction and Competence*. Development and Cell Biology Series 16. Cambridge: Cambridge University Press.

Nijhout, H. Frederik. 1991. *The Development and Evolution of Butterfly Wing Patterns*. Washington, DC: Smithsonian Institution Press.

Nijhout, H. Frederik. 2003. "Development and Evolution of Adaptive Polyphenisms." *Evolution & Development* 5 (1): 9–18.

Nilsson-Ehle, Herman. 1913. "Sur les travaux de sélection du Froment et de l'Avoine exécutés à Svalöf 1900-1912," *Bulletin mensuel des renseignements agricoles et des maladies des plantes* 4 (6): 861–70. https://vlp.mpiwg-berlin.mpg.de/library/data/lit29289/index_html?pn=3&ws=1.5.

Nilsson-Ehle, Herman. 1914. "Vilka erfarenheter hava hittills vunnits rörande möjligheten av växters acklimatisering?" In *Kungl. Landtbruks Akademiens Handlingar och Tidskrift* 22 (1): 537–72.

Normann, Claus, et al. 2007. "Long-Term Plasticity of Visually Evoked Potentials in Humans Is Altered in Major Depression." *Biological Psychiatry* 62 (5): 373–80.

Novoplansky, Ariel. 2002. "Developmental Plasticity in Plants: Implications of Non-cognitive Behavior." *Evolutionary Ecology* 16 (3): 177–88.

Nussey, D. H., A. J. Wilson, and Jean E. Brommer. 2007. "The Evolutionary Ecology of Individual Phenotypic Plasticity in Wild Populations." *Journal of Evolutionary Biology* 20 (3): 831–44.

Nylin, Sören, Perolof Wickman, and Christer Wiklund. 1989. "Seasonal Plasticity in Growth and Development of the Speckled Wood Butterfly, *Pararge aegeria* (Satyrinae)." *Biological Journal of the Linnean Society* 38 (2): 155–71.

Odling-Smee, F. John, Kevin N. Laland, and Marcus W. Feldman. (2003) 2013. *Niche Construction: The Neglected Process in Evolution (MPB-37)*. Princeton, NJ:. Princeton University Press.

Oppenheimer, Jane M. 1936. "Historical Introduction to the Study of Teleostean Development." *Osiris* 2: 124–48.

Oppenheimer, Jane M. 1955a. "Problems, Concepts and Their History [in Embryology]." In *Analysis of Development*, edited by B. H. Willier, P. A. Weiss, and V. Hamburger, 1–24. Philadelphia: Saunders.

Oppenheimer, Jane M. 1955b. "Methods and Techniques." In *Analysis of Development*, edited by B. H. Willier, P. A. Weiss, and V. Hamburger, 25–38. Philadelphia: Saunders.

Oppenheimer, Jane M. 1967. *Essays in the History of Embryology and Biology*. Cambridge, MA: MIT Press.

Osborn, Henry F. 1897. "The Limits of Organic Selection." *American Naturalist* 31 (371): 944–51.

Oyama, Susan. (1985) 2000. *The Ontogeny of Information: Developmental Systems and Evolution*. Durham, NC: Duke University Press.

Oyama, Susan. 2000a. "Causal Democracy and Causal Contributions in Developmental Systems Theory." *Philosophy of Science* 67 (S3): S332–47.

Oyama, Susan. 2000b. *Evolution's Eye: A Systems View of the Biology-Culture Divide*. Durham, NC: Duke University Press.

Oyama, Susan, Russell D. Gray, and Paul E. Griffiths, eds. 2003. *Cycles of Contingency: Developmental Systems and Evolution*. Cambridge, MA: MIT Press.

Pfeifer, Edward J. 1965. "The Genesis of American Neo-Lamarckism." *Isis* 56 (2): 156–67.

Pfennig, David W. 2004. "Developmental Plasticity and Evolution." *American Scientist* 92 (1): 84.

Pfennig, David W. 2021. *Phenotypic Plasticity and Evolution: Causes, Consequences, Controversies*. Taylor & Francis.

Piaget, Jean. 1950. *Introduction à l'épistémologie génétique : La pensée biologique, la pensée psychologique et la pensée sociologique*. [Introduction to genetic epistemology, biological thinking, psychological thinking and sociological thinking]. Paris: PUF.

Piaget, Jean. 1968. *Le structuralisme* [Structuralism]. Paris: Que sais-je?

Piaget, Jean, and Rolando Garcia. 1983. *Psychogenèse et histoire des sciences*. Paris: Flammarion.

Pigllucci, Massimo. 1996. "How Organisms Respond to Environmental Changes: From Phenotypes to Molecules (and Vice Versa)." *Trends in Ecology & Evolution* 11 (4): 168–73.

Pigliucci, Massimo. 2001. *Phenotypic Plasticity: Beyond Nature and Nurture*. Baltimore: John Hopkins University Press.

Pigliucci, Massimo, and Gerd B. Müller. 2010. *Evolution: the Extended Synthesis*. Cambridge, MA: MIT Press.

Plaisance, Kathryn S., and Thomas A. C. Reydon, eds. 2012. *The Philosophy of Behavioral Biology*. Dordrecht: Springer Netherlands,

Plotkin, Henry. (1993) 1997. *Darwin, Machines, and the Nature of Knowledge*. Cambridge, MA: Harvard University Press.

Prior, Elizabeth. 1985. *Dispositions*. Aberdeen: Aberdeen University Press.

Rádl, Emanuel. 1930. *The History of Biological Theories*. Oxford: Oxford University Press.

Raisman, Geoffrey. 1969. "Neuronal Plasticity in the Septal Nuclei of the Adult Rat." *Brain Research* 14 (1): 25–48.

Rathke, Heinrich. 1861. *Entwickelungsgeschichte der Wirbelthiere*. Leipzig: Engelmann.

Rehfeldt, Gerald E. 1979. "Ecotypic Differentiation in Populations of *Pinus monticola* in North Idaho: Myth or Reality?" *American Naturalist* 114 (5): 627–36.

Reid, Robert G. B. 2009. *Biological Emergences: Evolution by Natural Experiment*. Cambridge, MA: MIT Press.

Rendel, James M. 1967. *Canalization and Gene Control*. London: Academic Logos Press.

Rensch, Bernhard. (1947) 1966. *Evolution above the Species Level*. New York: John Wiley & Sons.

Resnik, David B. 1992. "Discussion: Leo Buss's *The Evolution of Individuality*." *Biology and Philosophy* 7, no. 4): 453–60.

Rice, Sean H. 1998. "The Evolution of Canalization and the Breaking of Von Baer's Laws: Modeling the Evolution of Development with Epistasis." *Evolution* 52 (3): 647–56.

Richards, Robert J. (2002) 2010. *The Romantic Conception of Life: Science and Philosophy in the Age of Goethe*. Chicago: University of Chicago Press.

Ridley, Mark. 1981. "De Beer, Gavin Rylands." In Charles Gillispie (ed) *Dictionary of Scientific Biography*, edited by David Resnik, 17:213–14. New York: Charles Scribner's Sons.

Ridley, Mark. 1986. "Embryology and Classical Zoology in Great Britain." In *A History of Embryology*, edited by T. J. Horder, J. H. Witkowski, and C. C. Wylie, 35–67. Cambridge: Cambridge University Press.

Rinkevich, Baruch, and Valeria Matranga. 2009. *Stem Cells in Marine Organisms*. The Netherlands: Springer.

Robert, Jason Scott, Brian K. Hall, and Wendy M. Olson. 2001. "Bridging the Gap between Developmental Systems Theory and Evolutionary Developmental Biology." *BioEssays* 23 (10): 954–62.

Roe, Shirley A. 1979. "Rationalism and Embryology: Caspar Friedrich Wolff's Theory of Epigenesis." *Journal of the History of Biology* 12:1–43.

Roe, Shirley A. (1981) 2003. *Matter, Life, and Generation: Eighteenth-Century Embryology and the Haller-Wolff Debate*. Cambridge: Cambridge University Press.

Roff, Derek A. (1997) 2012. *Evolutionary Quantitative Genetics*. Springer Science and Business Media.

Roger, Jacques. (1963) 2014. *Les sciences de la vie dans la pensée française au XVIIIe siècle: La génération des animaux de Descartes à l'Encyclopédie*. Paris: Albin Michel, 2014.

Rollo, C. David. 2004. "Life = Epigenetics, Ecology, and Evolution (L= E3): A Review of Developmental Plasticity and Evolution by Mary Jane West-Eberhard." *Evolution & Development* 6 (1): 58–62.

Romaschoff, D. D. 1925. "Über die Variabilität in der Manifestierung eines erblichen Merkmales (Abdomen abnormalis) bei *Drosophila funebris* F." *Journal für Psychologie und Neurologie* 31 (5): 323–25.

Rosenberg, Alexander. 1994. *Instrumental Biology, or the Disunity of Science*. Chicago: University of Chicago Press.

Rosenblatt, Jay S. 1995. *Daniel S. Lehrman, 1919–1972: A Biographical Memoir* Washington, DC: National Academies Press. https://www.nasonline.org/publications/biographical-memoirs/memoir-pdfs/lehrman-daniel-s.pdf.

Roth, V. Louise. 1993. "Dwarfism and Variability in the Santa Rosa Island Mammoth: An Interspecific Comparison of Limb Bone Sizes and Shapes in Elephants." In *Third California Islands Symposium: Recent Advances in Research on the California Islands*, edited by F. G. Hochberg, 433–42. Santa Barbara, CA: Santa Barbara Museum of Natural History.

Roux, Wilhelm. 1888. "Beiträge zur Entwickelungsmechanik des Embryo." *Archiv für pathologische Anatomie und Physiologie und für klinische Medicin* 114 (2): 246–91.

Roy, J., and H. A. Mooney. 1982. "Physiological Adaptation and Plasticity to Water Stress of Coastal and Desert Populations of *Heliotropium curassavicum* L." *Oecologia* 52 (3): 370–75.

Rutherford, Suzanne L., and Susan Lindquist. 1998. "Hsp90 as a Capacitor for Morphological Evolution." *Nature* 396 (6709): 336–42.

Sachs, Tsvi. 2002. "Consequences of the Inherent Developmental Plasticity of Organ and Tissue Relations." *Evolutionary Ecology* 16, no. 3): 243–65.

Saint-Hilaire, Isidore Geoffroy. 1837. *Histoire générale et particulière des anomalies de l'organisation chez l'homme et les animaux: Ouvrage comprenant des recherches sur les caractères, la classification, l'influence physiologique et pathologique, les rapports généraux, les lois et les causes des monstruosités, variétés et vices de conformation, ou traité de tératologie.* Vol. 1. Paris: Établissement Encyclographique.

Salazar-Ciudad, Isaac, Jukka Jernvall, and Stuart A. Newman. 2003. "Mechanisms of Pattern Formation in Development and Evolution." *Development* 130 (10): 2027–37.

Salmon, Wesley C. 1971. *Statistical Explanation and Statistical Relevance.* Vol. 69. Pittsburg, PA. University of Pittsburgh Press.Sapp, Jan. 1987. *Beyond the Gene: Cytoplasmic Inheritance and the Struggle for Authority in Genetics.* Oxford: Oxford University Press.

Sarich, Vincent M. 1980. "A Macromolecular Perspective on the Material Basis of Evolution." In *Richard Goldschmidt, Controversial Geneticist and Creative Biologist: A Critical Review of His Contributions with an Introduction by Karl von Frisch*, edited by Leonie K. Piternik, Experientia Supplementum 35, 27–31. Basel: Birkhäuser.

Sarkar, Sahotra. 1999. "From the *Reaktionsnorm* to the Adaptive Norm: The Norm of Reaction, 1909–1960." *Biology and Philosophy* 14: 235–52.

Sawyer, Roger H., and Loren W. Knapp. 2003. "Avian Skin Development and the Evolutionary Origin of Feathers." *Journal of Experimental Zoology Part B: Molecular and Developmental Evolution* 298 (1): 57–72.

Scarth, George William. 1927. "Stomatal Movement: Its Regulation and Regulatory Role; A Review." *Protoplasma* 2 (1): 498–511.

Schaffner, Kenneth F. 1998. "Genes, Behavior, and Developmental Emergentism: One Process, Indivisible?" *Philosophy of Science* 65 (2): 209–52.

Schaffner, Kenneth F. 2006. "Reduction: The Cheshire Cat Problem and a Return to Roots." *Synthese* 151 3): 377–402.

Scharloo, Willem. 1991. "Canalization: Genetic and Developmental Aspects." *Annual Review of Ecology and Systematics* 22, no. 1): 65–93.

Scheiner, Samuel M. 1993. "Genetics and Evolution of Phenotypic Plasticity." *Annual Review of Ecology and Systematics* 24, no. 1): 35–68.

Scheiner, Samuel M., and Richard F. Lyman. 1989. "The Genetics of Phenotypic Plasticity. I. Heritability." *Journal of Evolutionary Biology* 2 (2): 95–107.

Scheiner, Samuel M., and Richard F. Lyman. 1991. "The Genetics of Phenotypic Plasticity. II. Response to Selection." *Journal of Evolutionary Biology* 4 (1): 23–50.

Schlichting, Carl D. 1986. "The Evolution of Phenotypic Plasticity in Plants." *Annual Review of Ecology and Systematics* 17 (1): 667–93.

Schlichting, Carl D., and Massimo Pigliucci. 1993. "Control of Phenotypic Plasticity via Regulatory Genes." *American Naturalist* 142 (2): 366–70.

Schlichting, Carl D., and Massimo Pigliucci. 1998. *Phenotypic Evolution: A Reaction Norm Perspective.* Sunderland, MA: Sinauer.

Schlosser, Gerhard, and Günter P. Wagner, eds. 2004. *Modularity in Development and Evolution.* Chicago: University of Chicago Press.

Schmalhausen, Ivan Ivanovich. (1949) 1986. *Factors of Evolution: The Theory of Stabilizing Selection.* Chicago: University of Chicago Press.

Schmitt, Johanna, Alex C. McCormac, and Harry Smith. 1995. "A Test of the Adaptive Plasticity Hypothesis Using Transgenic and Mutant Plants Disabled in Phytochrome-Mediated Elongation Responses to Neighbors." *American Naturalist* 146 (6): 937–53.

Sedgwig, Alan. 1889. *Studies from the Morphological Laboratory in the University of Cambridge.* Vol. 4. London: Clay.

Shapiro, Arthur M. 1976. "Seasonal Polyphenism." *Evolutionary Biology* 9:259–333. Boston: Springer.

Shaw, Ruth G., Gerrit A. Platenkamp, Frank H. Shaw, and Robert H. Podolsky. 1995. "Quantitative Genetics of Response to Competitors in *Nemophila menziesii*: A Field Experiment." *Genetics* 139 (1)): 397–406.

Siegal, Mark L., and Aviv Bergman. 2002. "Waddington's Canalization Revisited: Developmental Stability and Evolution." *Proceedings of the National Academy of Sciences* 99 (16): 10528–32.

Simpson, George G. (1944) 1984. *Tempo and Mode in Evolution.* New York: Columbia University Press, 1984.

Simpson, George G. (1953) 1961. *The Major Features of Evolution.* New York: Columbia University Press.

Simpson, George G., and Laurence Simpson. (1949) 1967. *The Meaning of Evolution: A Study of the History of Life and of Its Significance for Man.* Vol. 23. New Haven, CT: Yale University Press.

Sinnott, Edmund W., Leslie. C. Dunn, and Theodosius A. Dobzhansky (1925) 1950. *The Principles of Genetics,* 4th ed. New York: McGraw-Hill.

Slatkin, Montgomery. 1987. "Quantitative Genetics of Heterochrony." *Evolution* 41 (4): 799–811.

Sloan, Phillip R. 2002. "Performing the Categories: Eighteenth-Century Generation Theory and the Biological Roots of Kant's A Priori." *Journal of the History of Philosophy* 40(2): 229–53.

Smith, Harry. 1990. "Signal Perception, Differential Expression within Multigene Families and the Molecular Basis of Phenotypic Plasticity." *Plant, Cell & Environment* 13 (7): 585–94.

Smith, John Maynard. 1990. "Models of a Dual Inheritance System." *Journal of Theoretical Biology* 143 (1): 41–53.

Smith-Gill, Sandra J. 1983. "Developmental Plasticity: Developmental Conversion versus Phenotypic Modulation." *American Zoologist* 23 (1): 47–55.

Sober, Elliott. 1980. "Evolution, Population Thinking, and Essentialism." *Philosophy of Science* 47, no. 3): 350–83.

Sober, Elliott. 1983. "Equilibrium Explanation." *Philosophical Studies: An International Journal for Philosophy in the Analytic Tradition* 43 (2): 201–10.

Sober, Elliott. (1984) 2014. *The Nature of Selection: Evolutionary Theory in Philosophical Focus.* Chicago: University of Chicago Press.

Society For Experimental Biology. 1948. *Growth: In Relation to Differenciation and Morphogenesis.* Cambridge: Published for the Company of Biologists.

Solmsen, Friedrich. 1963. "Nature as Craftsman in Greek Thought." *Journal of the History of Ideas* 24 (4): 473–96.

Spemann, Hans. 1935. "The Organizer-Effect in Embryonic Development." Nobel Prize speech, December 12, 1935. https://www.nobelprize.org/prizes/medicine/1935/spemann/lecture/.

Spemann, Hans. 1938. *Embryonic Development and Induction.* New Haven, CT: Yale University Press.

Spemann, Hans, and Hilde Mangold. (1923) 2003. "Induction of Embryonic Primordia by Implantation of Organizers from a Different Species, 1923." *International Journal of Developmental Biology* 45 (1): 13–38.

Stahl, Georg Ernst. 1737. *Theoria medica vera.*

Stearns, Stephen C. 1983. "Introduction to the Symposium: The Inter-Face of Life-History Evolution, Whole-Organism Ontogeny and Quantitative Genetics." *American Zoologist* 23 (3/4): 1–125.

Stearns, Stephen C. 1989. "The Evolutionary Significance of Phenotypic Plasticity." *Bioscience* 39 (7: 436–45.

Stebbins, G. Ledyard. 1950. *Variation and Evolution in Plants.* New York: Columbia University Press.

Stern, Curt. 1980. "Richard Benedict Goldschmidt (1878–1958): A Biographical Memoir." In *Richard Goldschmidt, Controversial Geneticist and Creative Biologist: A Critical Review of His Contributions with an Introduction by Karl von Frisch,* edited by Leonie K. Piternick, 68–99. Basel: Birkhäuser.

Tabery, James. 2014. *Beyond Versus: The Struggle to Understand the Interaction of Nature and Nurture.* Cambridge, MA: MIT Press.

Takagi, Hiroaki, and Kunihiko Kaneko. 2005. "Dynamical Systems Basis of Metamorphosis: Diversity and Plasticity of Cellular States in Reaction Diffusion Network." *Journal of Theoretical Biology* 234 (2): 173–86.

Thoday, John M. 1953. "Components of Fitness." In *Evolution.* Symposia of the Society for Experimental Biology 7, 96–113. New York: Academic Press.

Thoday, John M. 1958. "Homeostasis in a Selection Experiment." *Heredity* 12 (4): 401–15.

Timoféeff-Ressovsky, N. W. 1925. "Über den Einfluss des Genotypus auf das phänotypen Auftreten eines einzelnes Gens." *Journal für Psychologie und Neurologie* 31 (5): 305–10.

Timoféeff-Ressovsky, H. A., and N. W. Timoféeff-Ressovsky. 1926. "Über das phänotypische Manifestieren des Genotyps. II. Über idio-somatische Variationsgruppen bei *Drosophila funebris.*" *Wilhelm Roux'Archiv für Entwicklungsmechanik der Organismen* 108:146–70.

Van der Weele, Cor. 1999. *Images of Development: Environmental Causes in Ontogeny.* New York: State University of New York Press.

Van Tienderen, Peter H. 1991. "Evolution of Generalists and Specialists in Spatially Heterogeneous Environments." *Evolution* 45 (6): 1317–31.

Van Tienderen, Peter H., Ibtisam Hammad, and Frits C. Zwaal. 1996. "Pleiotropic Effects of Flowering Time Genes in the Annual Crucifer *Arabidopsis thaliana* (Brassicaceae)." *American Journal of Botany* 83 (2): 169–74.

Van Tienderen, Peter H., and Hans P. Koelewijn. 1994. "Selection on Reaction Norms, Genetic Correlations and Constraints." *Genetics Research* 64 (2): 115–25.

Via, Sara. 1984a. "The Quantitative Genetics of Polyphagy in an Insect Herbivore. I. Genotype-Environment Interaction in Larval Performance on Different Host Plant Species." *Evolution* 38 (4): 881–95.

Via, Sara. 1984b. "The Quantitative Genetics of Polyphagy in an Insect Herbivore. II. Genetic Correlations in Larval Performance within and among Host Plants." *Evolution* 38 (4): 896–905.

Via, Sara. 1987. "Genetic Constraints on the Evolution of Phenotypic Plasticity." *Genetic Constraints on Adaptive Evolution*, edited by V. Loeschcke 47–71. Berlin: Springer.

Via, Sara. 1993. "Adaptive Phenotypic Plasticity: Target or By-Product of Selection in a Variable Environment?" *American Naturalist* 142 (2): 352–65.

Via, Sara, Richard Gomulkiewicz, Gerdien De Jong, Samuel M. Scheiner, Carl D. Schlichting, and Peter H. Van Tienderen. 1995. "Adaptive Phenotypic Plasticity: Consensus and Controversy." *Trends in Ecology & Evolution* 10 (5): 212–17.

Via, Sara, and Russell Lande. 1985. "Genotype-Environment Interaction and the Evolution of Phenotypic Plasticity." *Evolution* 39 (3): 505–22.

Via, Sara, and Russell Lande. 1987. "Evolution of Genetic Variability in a Spatially Heterogeneous Environment: Effects of Genotype-Environment Interaction." *Genetics Research* 49 (2): 147–56.

Vogt, Oskar. 1926. "Psychiatrisch wichtige Tatsachen der zoologisch-botanischen Systematik." *Zeitschrift für die gesamte Neurologie und Psychiatrie* 101 (1): 805–32.

Von Dassow, George, and Ed Munro. 1999. "Modularity in Animal Development and Evolution: Elements of a Conceptual Framework for Evodevo." *Journal of Experimental Zoology* 285 (4): 307–25.

von Wolff, Christian. 1731. *Cosmologie Generalis: Methodo Scientifica Pertactata*. Frankfurt: Officina Libraria Rengeriana.

von Wolff, Christian. 1736. *Philosophia Prima Sive Ontologia, Methodo Scientifica Pertractata: Qua Omnis Cognitionis Humanae Principia Continentur*. Frankfurt: Officina Libraria Rengeriana.

Waagen, Wilhelm Heinrich. 1886. "Note on Some Palaeozoic Fossils Recently Collected by Dr. H. Warth in the Olive Group of the Salt-Range." *Records, Geological Society of India* 19 (1): 22–38. https://babel.hathitrust.org/cgi/pt?id=hvd.32044107307464&seq=6.

Waddington, Conrad Hal. (1938) 2016. *An Introduction to Modern Genetics*. London: Routledge.

Waddington, Conrad H. 1942. "Canalization of Development and the Inheritance of Acquired Characters." *Nature* 150 (3811): 563–65.

Waddington, Conrad H. 1953. "Genetic Assimilation of an Acquired Character." *Evolution* 7 (2): 118–26.

Waddington, Conrad Hal. (1957) 2014. *The Strategy of the Genes*. London: Routledge.

Waddington, Conrad H. 1959. "Canalization of Development and Genetic Assimilation of Acquired Characters." *Nature* 183, no. 4676): 1654–55.

Waddington, Conrad Hal. 1961. "Genetic Assimilation." *Advances in Genetics* 10:257–93.

Wagner, Andreas. (2007) 2013 *Robustness and Evolvability in Living Systems*. Vol. 24. Princeton, NJ: Princeton University Press.

Wagner, Günter P., and Lee Altenberg. 1996. "Perspective: Complex Adaptations and the Evolution of Evolvability." *Evolution* 50 (3): 967–76.

Wagner, Günter P., Ginger Booth, and Homayoun Bagheri-Chaichian. 1997. "A Population Genetic Theory of Canalization." *Evolution* 51 (2): 329–47.

Waisbren, Steven James. 1988. "The Importance of Morphology in the Evolutionary Synthesis as Demonstrated by the Contributions of the Oxford Group: Goodrich, Huxley, and De Beer." *Journal of the History of Biology* 21:291–330.

Wallace, Bruce. 1986. "Can Embryologists Contribute to an Understanding of Evolutionary Mechanisms?" *Integrating Scientific Disciplines*, edited by William Bechtel, 149–63. Dordrecht: Springer Netherlands.

Walsh, Denis M. 1998. "The Scope of Selection: Sober and Neander on What Natural Selection Explains." *Australasian Journal of Philosophy* 76 (2): 250–64.

Walsh, Denis M. 2015. *Organisms, Agency, and Evolution.* Cambridge: Cambridge University Press.

Weber, Marcel. 2004. *Philosophy of Experimental Biology.* Cambridge: Cambridge University Press.

Wessells, Normal K. 1977. *Tissue Interactions and Development.* Menlo Park, CA: Benjamin-Cummings .

West-Eberhard, Mary Jane. 1989. "Phenotypic Plasticity and the Origins of Diversity." *Annual Review of Ecology and Systematics* 20 (1): 249–78.

West-Eberhard, Mary Jane. 2003. *Developmental Plasticity and Evolution.* Oxford: Oxford University Press.

West-Eberhard, Mary Jane. 2005. "Phenotypic Accommodation: Adaptive Innovation due to Developmental Plasticity." *Journal of Experimental Zoology Part B: Molecular and Developmental Evolution* 304 (6): 610–18.

Wiener, Norbert. (1961) 2019. *Cybernetics or Control and Communication in the Animal and the Machine.* Cambridge, MA: MIT Press.

Wilkins, Adam S. 2002. *The Evolution of Developmental Pathways.* Sunderland, MA: Sinauer.

Williams, George C. (1966) 2018. *Adaptation and Natural Selection: A Critique of Some Current Evolutionary Thought.* Vol. 61. Princeton, NJ: Princeton University Press.

Wimsatt, William C. 2007. *Re-engineering Philosophy for Limited Beings: Piecewise Approximations to Reality.* Cambridge, MA: Harvard University Press.

Wimsatt, William C. 2014. "Entrenchment as a Theoretical Tool in Evolutionary Developmental Biology." In *Conceptual Change in Biology: Scientific and Philosophical Perspectives on Evolution and Development,* edited by Alan C. Love, 365–402. Dordrecht: Springer Netherlands.

Winther, Rasmus G. 2000. "Darwin on Variation and Heredity." *Journal of the History of Biology* 33): 425–55.

Wolff, Caspar F. 1759. *Theoria generationis.*

Wolff, Caspar F, and Jean-Claude Dupont, trans. 2003. *De formatione intestinorum: La Formation des Intestins (1768–1769).* Edited by Jean-Claude Dupont. Turnhout: Brepols.

Wolff, Caspar F., and T. A. Lukina, trans. 1973. *Objecta Meditationum pro Theoria Monstrorum.* Leningrad: Nauka.

Woltereck, Richard. 1909. "Weitere experimentelle Untersuchungen uber Artveranderung, speziell uberdas Wesen quantitativer Artunterschyiede bei Daphniden." *Verhandlungen der Deutschen zoologischen Gesellschaft* 19:110–72.

Woodger, Joseph H. (1929) 2014. *Biological Principles: A Critical Study.* London: Routledge.

Woolf, Virginia. 1928. *Orlando: A Biography.* New York. Harcourt, Brace.

Woodward, James, and Lauren Ross. "Scientific Explanation." 2021. *The Stanford Encyclopedia of Philosophy,* edited by E. N. Zalta and Uri Nodelman. https://plato.stanford.edu/archives/sum2021/entries/scientific-explanation/.

Wright, Sewall. 1920. "The Relative Importance of Heredity and Environment in Determining the Piebald Pattern of Guinea-Pigs." *Proceedings of the National Academy of Sciences* 6 (6): 320–32.

Wright, Sewall. 1949. "The Genetical Structure of Populations." *Annals of Eugenics* 15 (1): 323–54.

Index

Page numbers in italics refer to figures.

abstract ideas, 238. *See also* intuition, of life's metaphysical order

accidental benefit, 194. *See also* adaptation

accommodation: as adaptation, 196; definition of, 160; genetic, 218; phenotypic, 161, 218; in psychology, 195. *See also* adaptation; assimilation; organic selection

activation, 147–50; of lactose operon, 166. *See also* induction; inhibition

adaptation, 57, 59, 77, 82, 94, 108, 130, 199; versus "adaptive," 194, 199; definition of, 163, 182, 267n12; evolutionary, 130, 223; genetic, 77, 223; as habituation, 91; individual, 75, 218; and natural selection, 102, 175; norm of reaction as, 78; norm of reactivity as, 81; ontogenetic, 91, 193–94, 196, 199; open, 245; and organisms, 241, 266n3; origin of, 163; of phenotypic trait, 103, 163; physiological, 165, 240; process of, 160–61, 163, 175, 190; as regulation, 70, 161, 163, 165, 168; response as, 107; result of, 55; selection of, 218; self-, 75; through lens of evolution, 191–203. *See also* accommodation; characters: adaptive; mimicry; mutation: adaptive; norm of reaction: adaptive interpretation of; responses: adaptive

adaptationist program, 209, 267n4

adaptive plasticity. *See* plasticity: adaptive

Agassiz, Louis, 90, 95

Allard, Robert, 110

alternative phenotype, 120, 219–20, 222, 264n15. *See also* norm of reaction; polyphenism

animism, 30, 33, 42, 52, 54; and Darwin's terminology, 59; and meaning of plasticity, 42; and plastic- terms, 44. *See also* vitalism

ANOVA (analysis of variance), 104. *See also* variance

antireductionist positions, 136

Aristotle, 3–4, 9–12, 14–17, 22, 26, 29, 33–34, 45, 47, 232, 259n5, 259n9

artificial selection, 183, 186, 196, 198

Ashby, William, 165

assimilation: and accommodation, 160; definition of, 161–63; genetic, 78, 109, 161–63, 183, 191–92, 194; phenotypic, 121; in psychology, 163. *See also* accommodation; organic selection

autoréglage. See self-regulation

Babcock, Ernest, 86

Bachelard, Gaston, 248

Badyaev, Alexander, 121

Baldwin, James Mark, 74–75. *See also* Baldwin effect

Baldwin effect, 77–78. *See also* organic selection

Balfour, Francis M., 44

Barthez, Paul-Joseph, 32, 243, 268n6

Bateman, Ken, 183

Bateson, Gregory, 257

Bateson, William, 59, 72–73, 79, 101, 110

Beatty, John, 82

behavior: animal, 16, 213; and automata, 16; cell, 52; diversity of, 244; and environment, 262n5; explanation of, 244; human, 244–45; modifying (*see* accommodation; regulation: in functional studies); noncognitive, 242; social, 216, 223; studies on (*see* vitalism: methodological)

behavioral plasticity. *See* plasticity: behavioral

benchmark: regulation as, 155, 167–69, 171–72; regulation driven by, 164–66. *See also* regulation